THE INFLUENCE OF GENETICS ON CONTEMPORARY THINKING

LOGIC, EPISTEMOLOGY, AND THE UNITY OF SCIENCE

VOLUME 6

Logic, Epistemology, and the Unity of Science aims to reconsider the question of the unity of science in light of recent developments in logic. At present, no single logical, semantical or methodological framework dominates the philosophy of science. However, the editors of this series believe that formal techniques like, for example, independence friendly logic, dialogical logics, multimodal logics, game theoretic semantics and linear logics, have the potential to cast new light no basic issues in the discussion of the unity of science.

This series provides a venue where philosophers and logicians can apply specific technical insights to fundamental philosophical problems. While the series is open to a wide variety of perspectives, including the study and analysis of argumentation and the critical discussion of the relationship between logic and the philosophy of science, the aim is to provide an integrated picture of the scientific enterprise in all its diversity.

The Influence of Genetics on Contemporary Thinking

Edited by

Anne Fagot-Largeault

Collège de France, Paris, France

Shahid Rahman

Université de Lille 3, France

and

Juan Manuel Torres

Universidade de Lisboa, Portugal and Universidad Nacional del Sur, Bahía Blanca, Argentina

Springer

A C.I.P. Catalogue record for this book is available from the Library of Congress.

ISBN 978-1-4020-5663-5 (HB)
ISBN 978-1-4020-5664-2 (e-book)

Published by Springer,
P.O. Box 17, 3300 AA Dordrecht, The Netherlands.

www.springer.com

Printed on acid-free paper

Cover image: Adaptation of a Persian astrolabe (Brass, 1712–13), from the collection of the Museum of the History of Science, Oxford. Reproduced by permission.

Springer

TABLE OF CONTENTS

PREFACE

1. INTRODUCTION

Defined as the branch of biology dealing with heredity and the mechanics of gene transmission from one generation to the next, genetics first made its appearance on the scientific scene at the beginning of the 20th century when H. de Vries, C. Correns and E. von Tschermak re-discovered the extraordinary experimental work of Mendel. It was this re-discovery that in just a few years gave rise to the unprecedented and explosive development of a new experimental approach in biology which was to provide the foundations of our modern understanding of medical and evolutionary biological phenomena – a revolution, one might say. And yet it is only as of the 1970s that we speak of a "revolution of genetics", as if the birth of genetics itself decades earlier had little or no transcendence in the history of science. This seeming paradox may be explained by drawing a distinction between the revolution signifying the birth within the realm of *basic science* of the very concept of genetics at the beginning of last century and the "revolution of genetics" which we nowadays associate with the outstanding success of genetics as a methodology where hypotheses on gene functions are tested, especially in relation to *genetic engineering.* It is not really the traditional distinction between *pure* and *applied* science which is here at stake but rather a complex relation between two aspects of science in constant interchange, namely *theory* and *practice*, where one cannot be fully developed without the other. Indeed genetic engineering has become indispensable by the acquisition of fundamental knowledge on genetics. In some respects the initial emergence of genetics might be assimilated to the emergence of a paradigm, Mendel's laws becoming rapidly and firmly embraced by leading members of the scientific community to replace former paradigms by an interweaving of controversies within several sciences searching to elucidate problems old and new.

The interplay of these two aspects of science and research gives us insight into the capital importance of genetics today. Genetics as both theory and praxis is omnipresent in contemporary thinking; there are few, if any, disciplines where genetics does not exert its influence in one way or another. It is this increasing pertinence of genetics to the sciences that explains the inclusion of this volume in the series *Logic, Epistemology and the Unity of Science:* a common factor in science becomes *ipso facto* a factor of unity.

Genetics may not be of direct relevance to all fields of science, and yet for a number of reasons the concept of genetics is at the centre of much of today's scientific discussion and social debate. Genetics has become the touchstone for resolving many of the controversies in basic research dealing with the nature, history, and evolution of life, as well as the origin of man and ethnic groups, providing us with the tools that have led to some of the most astonishing advances in determining the structure and function of living entities and being able to modify organisms. Genetically engineered products, substantial alterations in plants, animals and microorganisms, and the manufacture of vital pharmaceutical products all stem from the same source. The domino effect of these transformations is likely to bring about even more dramatic changes in our lives as time passes, triggering more heated debate on the ecological, economic, and social impacts of this potential.

The availability of genomic information and in particular the completion of the *Human Genome Project* in 2003 brings with it consequences that we can only begin to imagine. The very same principle that teaches us how human genes instruct cells to build entities and regulate organic processes also enables us to identify species, organisms and individuals, measure evolutionary distances, and ultimately comprehend the limits of our own physical existence.

One of the most outstanding consequences of contemporary genetics and its practical implications is found in the medical sciences, where genetic testing for hereditary disorders and disease risk is a growing reality. As with the genetic modification of living beings, these new techniques for detecting potential health problems and the accompanying therapies developed for treating genetic diseases have consequences that go far beyond their initial focus, impacting on health and health-care theory and policy and even generating new concepts and chapters in the world of social science, jurisprudence and economics.

The far-reaching impact of genetics thus goes beyond the natural, medical and social sciences: contemporary genetics also adds a new dimension to classical topics in philosophy, in particular epistemology and ethics. Epistemological doctrines dealing with the structure of theories, reduction, and the status of scientific concepts encounter in genetics a new field of application. And from the ethical viewpoint, the spectre of a future society dominated by eugenic practices and the design of generic offspring confronts practical philosophy with an unprecedented challenge. In the light of such a scenario it is no exaggeration to say that contemporary genetics is to philosophy what physics was in the 17th century to theology and metaphysics, with Newton's breakthrough in modern thinking.

The Influence of Genetics on Contemporary Thinking reflects how genetics has transcended its original boundaries to become a key branch of science to crucial issues across a wide palette of disciplines. Chapter by chapter it shows how genetics forms the basis of modern-day biology, making the volume especially attractive for scholars interested in learning about the great scientific developments of the last and current centuries.

Following a suggestion of Philippe Huneman (IHPST-Paris), who provided an extremely competent review of an early draft of the manuscript, we have divided the book into chapters grouped according to the following main topics:

- **Part I Genetics and the Life Sciences**
- **Part II Genetics and Philosophy of Science: The Reductionism Debate and Beyond**
- **Part III Genetics and the Ethical, Legal and Sociological Debate**

An interview of François Jacob by Anne Fagot-Largeault precede these chapters.

Let us now present a brief overview of the contents of each chapter.

OVERVIEW OF CONTENTS

INTERVIEW OF FRANÇOIS JACOB BY ANNE FAGOT-LARGEAULT

Following the development of the questions formulated by Anne Fagot-Largeault the *English* version of the interview of François Jacob, who shared in 1965 the Nobel Prize in physiology or medicine with André Lwoff and Jacques Monod, has been divided in three sections, namely: *The methodological and epistemological question*, *Do you wish to know when and how you will die?* and *Genetics and the ethical debate*. The contents of these sections of the interview already indicate the main topics of the book.

We would like to point out, that after the reading of the written version of the original source in French, some paragraphs have been revised. In order to leave the original interview intact we implemented the revisions in the English version only.

PART I GENETICS AND THE LIFE SCIENCES

The main objective of the paper by F. J. Ayala, M. A. Capó, M. Nadal and C. J. Cela-Conde (Chapter 1) **Genetics and the Human Lineage: Can Genetics Throw Some Light on the Evolution of the Human Lineage?** is to provide an overview of molecular phylogenetic contributions to the study of hominid evolution. It reviews and takes a stand on traditional debates such as the African origin of man, the evolutionary relationship between the great ape and human lineages and the size of the original human populations.

The authors point out that genetic data has to be seen as a constraint upon paleoanthropological research. This approach allows them to elucidate the discussion on the hypothesis of the Mitochondrial Eve, which states that all women descend from a single or very few women. Indeed, the authors show that this hypothesis stems from the confusion of genealogy of genes and genealogy of individuals. The discussion in paragraph two about extra and intra group diversity will certainly stimulate debates on the notion of races and connects to issues in rights and politics such as those discussed in Part III of the present book.

Finally, the paper shows how genetic discoveries have contributed to unravelling the phylogenetic history of distinctive human traits such as language and the masticatory apparatus.

In Chapter 2 **Genetics and Neuroscience: Some Examples of their Recent Convergence and of the Continuing Nature–Nurture Controversy, With**

Emphasis on Sleep Physiology C. Debru discusses the introduction of genetical concepts and viewpoints in neurobiological research. He describes mainly the studies of memory performed by Eric Kandel who discovered the genetical functionings involved in memory, and the neurophysiology of paradoxial sleep developed by Michel Jouvet. He traces the history of the hypotheses of Michel Jouvet of genetic reprogramming during this type of sleep, confronting it with recent evidence. More precisely, sleep physiology and pathology is discussed with special emphasis on Michel Jouvet's theory, which insists on the potential regulatory function of paradoxical sleep on the interaction between genetic and epigenetic features, and which, according to the author, provides a possible answer to the nature–nurture controversy.

The reader might find it fruitful to confront the author's reference to "epigenetic influence" with the discussion of this notion in Chapter 8 by L. van Speybroeck, G. van de Vijver and D. De Waele.

Chapter 3 **Who Made Genetic Codes, How and by What?** by K. Matsuno. This paper discusses genetics from a newly emerging point of view in science called *internalism.* Internalism has some antecedents in phenomenology, the thinking of J. J. von Uexküll, and the autopoiesis model of Maturana and Varela. Current major thinkers include Koichiro Matsuno, and Yukio-Pegio Gunji, Otto Rössler, and Stanley N. Salthe. Salthe's helpful overview of internalism[1] states that internalism becomes necessary if we try to make a science which begins with the fact that we are inside, as participants in, the universe that we are studying. Internalism applies to such advanced technological situations as cosmological knowledge in the face of the finite speed of light (we cannot get outside the universe, or see it whole) and operationalism, as well as to the situation of a newborn infant trying to manage in the world.

Matsuno has been trying to examine the most reduced internalist situations possible in a search for principles. Stated in other words, the internalist project can be viewed as an attempt to understand a system from within, with the inquirer being a part, inside the system, and therefore unable to see itself as if from outside. In contrast, the mirror would symbolize the stance taken up in standard (externalist) scientific modelling, delivering a spatiotemporally global picture of a whole system, describable in the universal present tense (as in: "organisms reproduce" or "a star's energy dissipates"). While externally we might describe, for example, a dinner – the setting, menu, and so on – internally the experience of the dinner is its sequence of tastes. The main reason for taking this stance is that generativity cannot be approached externally. The reader might recall Gilbert Ryle's distinction between *knowing how* and *knowing that,* which in this case seems to correspond

[1] Cf. Salthe, 'Internalism summarized' in Salthe's homepage. See Van De Vijver G, Stanley N, Salthe and Manuela Delpos (1998) (eds) Evolutionary systems: biological and epistemological perspectives on selection and self-organization. Kluwer, Dordrecht.

to the distinction between the internal and the external view of the internalists.[2] More generally, internalism "in science" seems to complement with the *dialogical perspective* "in philosophy" as developed by Kuno Lorenz, who has a fine-grained and dynamic approach to the distinction between the *me* and the *you* perspective and combines it with Ryle's two types of knowledge.[3] Moreover, Rahman and collaborators extended the dialogical approach to a general framework for the study of non-classical logics, some of which seem to be akin to the internalist approach to the logic of science.[4] However, internalists and dialogicians did not pool yet their results in a common project which might yield a fruitful future cooperation.

What, then, is internalism in science? At present it is the attempt to model a system as if from inside. For some internalists such as Matsuno the situation within a quantum wave function presents a paradigmatic case where science encounters internalism. Once more here we would like to suggest another link which deserves to be explored, namely internalism in quantum physics and the recent dynamical approach of Jaakko Hintikka to the notion of information-independence in quantum mechanics.[5]

In this paper K. Matsuno applies this internalist view on quantum physics in order to address the question of the origin of the genetic code.

Chapter 4 **Genetics, Life and Death: Genetics as Providing a Definition of Life and Death** by M. Morange. This paper describes how genetics first established and then weakened the link between genetics and a definition of life. M. Morange

---

[2] For a thorough discussion of the epistemological consequences of this distinction see Rao N (1994), A Semiotic Reconstruction of Ryle's Critique of Cartesianism. De Gruyter, Berlin-New York. Ryle's distinction has been invoqued by Rahman and Symons as one of the philosophical tenets behind the project of the series Logic, Epistemology and the Unity of Science (Rahman/Symons (2004) Logic, Epistemology and the Unity of Science: an Encyclopedic Project in the Spirit of Neurath and Diderot. In: Rahman S, Symons J, Gabbay D, van Bendegem JP (eds) Logic, epistemology and the unity of science. Kluwer, Dordrecht (LEUS, vol 1) 3–15.

[3] See Lorenz K (2005) Pragmatic and semiotic prerequisites for predication. A dialogue model. In: Vanderveken D (ed) Logic, thought and action. Springer, Dordrecht (LEUS, vol. 2) 343–358; Lorenz K (2006) Logic as tool of science versus logic as a scientific subject. In: van Benthem J, Heinzmann G, Rebusqui M, Visser H (eds) The age of alternative logics.Kluwer, Dordrecht (LEUS, vol 3) Chapter 19.

[4] See Rahman S, Keiff L (2005) On how to be a dialogician. In: Vanderveken D (ed) Logic, thought and action. Springer, Dordrecht (LEUS, vol 2), 359–408; Rahman S (2006) Non-normal dialogics for a wonderful world and more. In: van Benthem J, Heinzmann G, Rebusqui M, Visser H (eds) The age of alternative logics. Kluwer, Dordrecht (LEUS, vol 3) Chapter 20 .

[5] For an overview see Kolak D, Symons J (2004) (eds) Quantifiers, questions and quantum Physics. Essays on the philosophy of Jaakko Hintikka. Springer, Dordrecht . The dynamical approach of Hintikka is strongly connected to the dialogical approach both historically and conceptually (see Rahman S, Dégremont C (2006) The beetle in the box: exploring if-dialogues. In: Aho T, Pietarinen A-V (eds) Truth and Games. Helsinki: acta philosophica fennica, vol 78). The dynamic approach to quantum physics is connected too to linear logic, see Coecke B, Moore DJ, Smets S (2004) Logic of dynamics and dynamics of logic: some paradigm examples. In: Rahman S, Symons J, Gabbay D, van Bendegem JP (eds) Logic, epistemology and the unity of science. Kluwer, Dordrecht (LEUS, vol 1) 527–556.

discusses the purported role of this link in biology by properly stressing its essential role in the definition of evolution by natural selection.

The author stresses the point that the link between life and death that has remained a constant of philosophical investigations from Aristotle to Bichat has also kept its place in the genetic approach to life. Definitions of life through both molecular and population genetics have tried – with less success for the former – to justify the place of death in the economy of nature. Furthermore, the author distinguishes the evolutionary accounts of the phenomenon of death from other phenomena such as cell death. He argues that as findings accumulated on cell death, it appeared less and less linked with the program of development and more and more related to the capacities of organisms to adapt to changing environments. It is the absence of cell death, not its occurrence, which in most cases constitutes a threat to the organism. This result is part of the chain of arguments which support the conclusion of the author of the paper, namely that models, derived from genetics, can be used fruitfully to describe the transmission of behaviors and beliefs between humans. However, according to the author, these models are Darwinian. *They do not refer to the genes of geneticists, but somehow apply the same corpuscular model behind the success of genetics to the transmission of mental states.* Actually, as remarked by Philippe Huneman, the notion of *transmission* mentioned by Morange should be thought more generally as including not only mental states but all kinds of abstract items and constructions (memes) such as traditions.

PART II GENETICS AND PHILOSOPHY OF SCIENCE: THE REDUCTIONISM DEBATE AND BEYOND

The paper by F. Bouchard. **Moving Beyond the Influence of Molecular Genetics on the Debate About Reductionism in Philosophy of Biology** (Chapter 5), launches the discussion on reductionism which will also be addressed too by the other papers in this part of the book. This paper in particular addresses the issue of the reduction of genetics to molecular biology providing an elucidating and detailed account of this traditional question in philosophy of science. In the last section the author displays his main contribution which boils down to proposing to introduce into the debate the consideration of ecology conceived in an energy-oriented viewpoint.

According to Bouchard, molecular genetics inspired many philosophers to think about the very small to understand biological processes, but, according to the author, these times are over if the aim is to bring biology closer to chemistry and physics.

The reader might find it fruitful to confront the notion of "explanatory reduction" of Bouchard with the "token-token" reduction discussed by Jean Gayon in the following chapter.

In Chapter 6 **The Concept of Gene in Contemporary Biology**, J. Gayon addresses the challenging question as to whether the concept of *gene* is or is not a theoretical concept. The proposed answer is negative, which of course has large and important consequences – one of them being the issue on reductionism discussed by Bouchard on Chapter 5.

The paper makes several provocative claims supported by convincing arguments: first, that molecular biology *is largely decoupled from heredity*, which was the central concern of classical genetics. This point, often overlooked, has huge bearings on any philosophical attempt to reduce classical genetics to molecular biology. Hence the conclusion of the author, that the concept of gene belongs to genetics rather than to molecular biology.

According to Gayon genes should be conceived of as processes rather than structures and this drives the author to reconsider the case of reductionism. It is an interesting fact that the token/token reduction to which the reductive programs for genetics seem to be restricted according to Gayon appears to be quite akin to Bouchard's "explanatory reduction".

A new and stimulating view on the concept of gene emerges from this paper: the concept of gene should be conceived, according to this view, not as an object-based notion but as a dynamical notion.

In Chapter 7 **The Influence of Genetics on Philosophy of Science**, P. Lorenzano deals with the topic of reductionism, as Bouchard and Gayon do in their respective papers. However, this author writes in the framework of the *structuralist theory of science*, which advocates a semantic approach on philosophy of science.

The main argument of the structuralists is that the *Received View* that scientific theories are best conceived of as interpreted sets of axioms and derived theorems is inferior to the model-theoretical approach. Scientific theories are conceived by the structuralists as extra-linguistic entities[6] and the author of this paper applies this model-theoretical approach to genetics. The reader might recall that there are some important antecedents of such reconstructions in Biology in the closely related approach called the *semantic view*.[7] It should be noticed that the issue at stake in this chapter is *classical genetics*. In fact, the reconstruction of genetics developed in this chapter relates to genetics before the irruption of molecular biology, contrary to the papers of Bouchard and Gayon, and more generally to the other papers of the volume.

The author convincingly shows that the fundamental law vindicated, *the law of matching*, makes sense of Mendel's laws, although as remarked by Philippe Huneman, before extending this result to molecular biology a careful examination of the assumptions involved is required, for example in the case of genes that have no organismic phenotypes, such as certain segregation distorters.

Chapter 8 **Epi-Geneticization: Where Biological and Philosophical Thinking Meet** by L. Van Speybroeck, G. Van de Vijver and D. De Waele. Philosophy of

[6] Cf. Balzer W, Mulines U, Sneed JD (1987) An architechtonic for science. Kluwer, Dordrecht.

[7] Cf. Beatty J (1980) Optimal-Design and the Strategy of Model Building in Evolutionary Biology, *Philos Sci* 47: 532–561, Batty J (1981) What's Wrong with the Received View of Evolutionary Theory? In: Asquith PD, Nickles T (eds), *PSA 1980*, Philosophy of Science Association, vol. 2, East Lansing, Michigan, 397–426, Lloyd E (1988) *The Structure and Confirmation of Evolutionary Theory*. Greenwood Press, New York, Thompson P (1989) The structure of biological theories. SUP-New York, Albany.

biology has a long history of taking classical and molecular genetics under analysis, and has often described genetics as gene-centric and reductionist. The authors of this chapter argue that this analysis has become outdated and that today "geneticization" should be interpreted as "epi-geneticization". This conceptual shift is supported by experimental research in molecular biology itself, showing that molecular biology is already taking up the challenge of approaching biology in less gene-centric terms. The authors argue that the question of epigenetics relates to the question of the extension of genic contexts. Furthermore, with the help of the notion of epigenetics they establish that there is a shift between genes as explanans towards genes as explanandum.

This statement might be confronted with Gayon's claim that the gene is no longer a theoretical concept but rather a *conceptual currency*. It might well be that the views of the present chapter and those of Gayon develop two complementary approaches to the contemporary state of genetics and molecular biology.

PART III GENETICS AND THE ETHICAL, LEGAL AND SOCIOLOGICAL DEBATE

The paper by A. Fagot-Largeault **Is DNA Revolutionizing Medicine?** (Chapter 9) launches the series of papers in this part of the book on the ethical, legal and sociological debate on the use of genetics in medicine. Actually Fagot-Largeault places the debate at the very start: did the discovery of the DNA molecule's double helix really revolutionize medicine? As far as clinicians can see, up to now it looks as if molecular genetics has had little impact on medical practice, and the question is still asked using the future tense: "Will genetics revolutionise medicine?" Furthermore and more generally, the author poses, and answers, central questions in relation to the present state of medicine regarding molecular genetics, by considering the view of the clinicians, philosophical issues about the theses of "geneticization" and the case of assisted procreation. The author describes the problems rigorously and argues in favor of prudent genetic intervention.

The author starts by describing the well-known general objection towards the use of new technologies in modern medicine based on the fear of dehumanization of clinical practice. Fagot-Largeault then pinpoints several precise examples of the use of genetics in medicine, aiming at a thorough response to the general objection mentioned above. One of the many convincing examples she presents in the paper is that of the use of a genetic screen to rule out unnecessary chemical therapy in the treatment of certain kinds of breast cancer. She concludes that *small signs like these give us a glimpse of how genetic technologies can be introduced into clinical practice, for the benefit of patients, and without leading to any major anthropological collapse.*

On Chapter 10 **The Harm of Being a Clone**, J.-Y.Goffi discusses the argument about the *harm* a clone would suffer in the case of a future secure device for implementing human reproductive cloning. This argument is currently used by the opponents to human cloning and the author explores rigorously the notion of harm involved. The author concludes that *a clone would not suffer any ontological or existential harm; but it would certainly suffer some psychological distress, resulting from irrational expectations on it.*

The path followed by the author to drive home this conclusion engages him in a deep philosophical debate about the harm inflicted to someone in the case that he has been brought to life "in order to fulfil someone else's expectations". The question is certainly difficult because, as the author concedes, this kind of harm can be inflicted on anyone, whether they are a clone or not.

An interesting fact is that one of the arguments discussed in the paper is the argument of the *right to ignorance* of H. Jonas. Jonas claims that cloning would violate a right to ignorance by failing to *respect the right of each human life to find its own way and be a surprise to itself.* In the interview which launches the book, while discussing the implementation of genetic diagnosis methods, F. Jacob defends the *right to knowledge*.

Chapter 11 **Children of One's Own** by J. M. Kaplan. This paper displays a perspective on the influence of genetics that is quite different to all the other papers of the volume, namely, the negative influence of some kind of rhetoric associated with the Human Genome Project on a crucial notion of our society: the notion of parenthood.

Indeed, the author challenges the link between genetic child and *child of one's own* and more generally he challenges the genetic notion of parenthood and makes a case for regarding the social relationships as defining parenthood. The author invokes several legal cases as evidence of the genetic-oriented ideology of parenthood and argues for another view of parenthood.

Kaplan starts challenging the overall presentation of technologies for assisted reproduction which often aim explicitly at giving hitherto infertile couples a *child of their own* – that is, a child that is genetically related to them. Kaplan formulates his main objection in the form of the following question: why should a genetic relationship make a child any more 'one's own' than other kinds of relationships – for example, those parent-child relationships forged through adoptions? Indeed, there seems to be a wide-spread assumption in much of contemporary society that genetic parenthood is important because of what it implies about the relationship between the (physical and behavioral) traits of the parents and those traits of the child; arguments relying on these assumptions have even been accepted in some legal cases. Kaplan argues in his incisive paper that this state of affairs is particularly unfortunate, and that the over-blown rhetoric of the Human Genome Project and related research programs is at least partially to blame. This rhetoric includes the metaphorical language of genes as "master controllers", "blue-prints", "recipes", and as "carrying information". He concludes that as none of these metaphors is well-justified by contemporary understandings of the roles played by genes in the organismal development, the metaphors ought to be rejected, and with them, the social emphasis on a genetic relationship as the most important aspect of parenthood.

In Chapter 12 **Is a Transcultural Law Possible?** C. M. Romeo Casabona assesses the state of the field concerning laws about genetics and gene research and cloning.

It explores the foundation of what could be a universal law in those matters arguing for a transcultural approach. He stresses some principles, added to the principles of human rights that could support this task: the principle of responsibility (Jonas), of solidarity (Sulston), of justice (Rawls), of equity and tolerance (Kaufmann and Saada-Gendron), the principle of non-discrimination and responsibility towards future generations (Casabona).

In his conclusion Casabona writes that international law has promoted a global perspective in relation to biomedical technology. Moreover, the author argues that this global perspective has been favored because the State laws lacked ethical and cultural reference points for a clear and undisputed application to the new challenges created by biomedical technology. He declares the need to go further to reach a transcultural ethics and law in the field of human genetics and biotechnology, that is to say, taking into consideration the significant contributions of the UNESCO Declaration on the Human Genome and Human Rights and the Convention on Human Rights and Biomedicine.

In Chapter 13 **Genetics and Society: A Different View**, J. M. Torres provides an overview on the current debates on health and genetics and emphasizes the Kuhnian revolution introduced by genetics in the theories of health. In this paper Torres explores the connection between genetics and health in a way that differs from other approaches, which usually consider the relations between genetics and society in terms of eugenic policies or those of the geneticization process, by focusing on the impact that genetic technologies have on the notions of health and unhealth. The author makes the case that health and disease should be separated because genetic tests have made it clear that one can be unhealthy without having a disease. The author is led to define health in terms of *basic biological requirements*. It consists of the possibility of going through the three stages that characterize human life and the lives of certain other species: childhood, adulthood, and old age. Torres argues that it is on the basis of this requirement that we usually project our future because, as intelligent beings, we have the capacity – at least in principle – to determine our own destiny by making decisions; typically by assuming long-term responsibilities such as getting married or having children. It should be clear that this basic biological requirement is not equivalent to health nor does it constitute its definition. Simply, this requirement is an obviously necessary condition for it.

As a final remark we recommend as further complementary reading Volume 4 of this series: *Nature's Principles*, edited by Jan Faye, Paul Needham, Uwe Scheffler and Max Urchs, which discusses in the context of general philosophy of science many of the issues dealt with in the present volume, such as reductionism and the very concept of scientific law.[8]

---

[8] Faye J, Needham P, Scheffler U, Urchs M (2005) (eds) Nature's principles. Springer, Dordrecht, (LEUS, vol 5).

AN HISTORICAL OUTLINE AND FURTHER READING

Genetics did not emerge fully armed, like Athena from the head of Zeus. It is rather the result of a long and still growing development. A development that has brought scientists and humanity from pure speculations about the roots of heredity to the possession of tools with which we today dominate many life processes.

Though *genetics* had its origins in the experiments carried out by the Augustinian prior Georg Mendel (1822–1884), it received its name from the British biologist William Bateson. Indeed, William Bateson was the first to suggest the word *genetics* (from the Greek *genno*, $\gamma\varepsilon\nu\omega$; *to give birth*) to describe the study of inheritance and the science of variation in a personal letter dated April 18, 1905. Bateson first used the term "genetics" publicly at the *Third International Conference on Genetics in London* in 1906, three years before the Danish Botanist Wilhelm Johannsen (1857–1927) used the word "gene" to describe the units of hereditary information.

Mendel's work was directed at shedding light on the patterns according to which hereditary characteristics are transmitted in organisms. From his experiments with peas and by very accurately employing artificial breeding techniques, he clearly succeeded where others had failed, an achievement imputable, to a great extent, to his impeccable use of scientific methods (Mendel, 1866; Dunn, 1965).

The so-called *Mendel's laws*, though they were not explicitly formulated by him, generalized his findings and let scientists glimpse the presence of true biological regularities. In this way, biology matched, at least partially, the traditional concept of science, understood as lawfulness knowledge. Thus, life sciences transcended for the first time the descriptive status, in the way that had Physics begun in the 17th century and Chemistry in the 19th century.

As with so many other cases in the history of science, Mendel's transcendental achievements did not obtain the recognition of the scientific community. In fact, though Mendel's results were not completely unknown to biologists of the time, they were not seen as being important. Even Mendel himself did not see their ultimate applicability, and thought they only applied to certain categories of species. In 1900, however, the work was rediscovered and re-evaluated.

After the re-discovery of Mendel the scientific and philosophical world was soon attracted to that type of entities which Mendel made responsible for the hereditary characteristics he studied, that he called *factors* and that we, since Johanssen, call *genes*. The great merit of the eminent American biologist Thomas Hunt Morgan (1866–1945), who endowed genetics with the character of autonomous discipline that it exhibits today, was to realize the crucial role that genetic mutations have in the evolutionary process, that is as being the true factor responsible for changes organic characteristics. Following the rediscovery of Mendelian inheritance in 1900, Morgan's research moved to the study of mutation in the fruit fly *Drosophila melanogaster*. In his famous *Fly Room* at Columbia University Morgan he was able to demonstrate that genes are carried on chromosomes and that they are the mechanical basis of heredity. In fact, when the Mendelian inheritance laws were integrated with Morgan's chromosome theory of inheritance the core of modern *classical genetics* was set. Morgan was awarded the Nobel Prize in Physiology or Medicine in 1933, the first person to receive this Prize for genetics.

The following years witnessed efforts to find out whether genes were proteins or nucleic acids, the two kinds of macromolecules that characterize living beings and are encountered in chromosomes. Finally, it was the bacteriologists Alfred Day Hershey (1908–1997), Nobel Prize in Physiology or Medicine in 1969, and Martha Cowles Chase (1927–2003) who in a famous experiment of 1952 went on to decide the debate in favor of nucleic acid (DNA), confirming the pioneer findings in 1944 of the Canadian physician Oswald Theodore Avery (1877–1955) and his colleagues, which had passed unnoticed by many in the scientific community.

The last step in this period of localization and identification of the hereditary substance was the discovery of American biologist James Dewey Watson (1928–) and the British physicists Francis Compton Crick (1916–2004), Rosalind Franklin (1920–1958) and Maurice Hugh Wilkins (1916–2004) of the DNA structure, also called since then *the molecule of heredity*. In their famous article of 1953 (a, b), probably the most frequently quoted paper in the history of science, Watson and Crick also suggested which are the mechanisms regulating the replication of the DNA and, therefore, the replication of genes, an hypothesis that was confirmed soon after. Crick, Watson and Wilkins were awarded the Nobel Prize in Physiology or Medicine in 1962.

However, the advances described, though important, were not enough to explain the relationship between the molecules of heredity and organic characteristics, that is, between genes and proteins. The problem to be solved was two-fold: firstly, how do genes *know* when they have to instruct cell machinery to produce the proteins that the organism demands, and how do they know when to begin and stop the enzyme production? Secondly, how are proteins *contained*, speaking informally, in the genes? It was the creative power of two French researchers, François Jacob (1920–) and Jacques Monod (1910–1976) that opened the door to begin solving the first mystery, at least for elementary organisms (the bacterium *E. coli*). With the earlier determination of the structure and central importance of DNA, it became clear that all proteins were being produced in some way from its genetic code, and that this step might form a key control point. In 1961 Jacob and Monod proposed a system, *the Operon-lac model*, through which it finally became possible to understand how genetic control of the protein synthesis was possible. They made key experimental and theoretical discoveries that showed that in the case studied there are specific proteins that are devoted to repressing the transcription of the DNA to its product (which is in turn decoded into a protein). Their experiments and ideas gave impetus to the emerging field of molecular development biology, and of transcriptional regulation in particular. For this achievement both were awarded the Nobel Prize in 1965, together with André Lwoff (1902–1994). Posterior works by Jacob and Monod with the collaboration of Jean-Pierre Changeux (1936–), enlarged our knowledge of the cellular mechanisms. In a nutshell: their monumental contributions showed that cells could be conceived as cybernetic systems whose functions are based in an amazing and harmonic interplay between genes and proteins. An interesting fact is that from the point of view implicit in Jacob's epistemological conceptions the emergence of a new paradigm might be placed before Mendel's work, namely around 1840 when the cellular theory was developed. The cellular theory, which says that all living beings are made of cells and that all living beings originate from a cell, allowed researchers in biology to start thinking *analytically*, as Jacob says. Cells are the basic units, they are observable, the development of a (human or other living) being can be followed and analyzed from the first cell on, and the biology of development may be considered as important achievement of the 20th century as the biology of evolution was an achievement of the 19th century. From that perspective, the point of molecular biology is precisely to analyse the machinery of the cell.

At that time, a time of extraordinary advances in life sciences, the American geneticist Marshall Warren Nirenberg (1927–) and the German biochemist Heinrich Matthaei (1929–), in the wake of the Spanish biochemist Severo Ochoa (1905–1993) who was awarded the Nobel Prize in Physiology or Medicine in 1959, solved the second mystery, that is the way in which proteins are *contained* in genes. Though the scientific community presumed that proteins should be within genes only in a virtual way, as instructions to carry out, details were unknown. The answer was given in 1961 by Nirenberg and Matthaei: specific nucleotide-sequences determine specific aminoacid-sequences. Therefore, the gene could be seen since then as a set of instructions written in the DNA language for building proteins, a language which is read and interpreted by the cellular machinery. The work of Ochoa, Nirenberg and Matthaei was actually the result of an international collaboration, whereby a particularly important role was played by the cooperation of the Indian biochemist Har Gobind Khorana (1922–) and of the American biochemist Robert W. Holley (1922–1993), who in 1968 shared the Nobel Prize in Physiology or Medicine with Nirenberg.

Now, given a set of instructions for performing intelligent work, could this set be conceived as a naked piece of raw material? Hardly so. The reason should be clear: it has an aggregate value, namely: meaning. In the case of genes, the meaning is directed to the cell that *knows* how to read the message encoding the instructions. Many philosophers saw in the differentiation between mere pieces of DNA and genes, understood as units of information, a reason to re-introduce the much criticized distinction between theoretical and observational terms (cf. Ruse, 1973). In a parallel way, if the reduction of the units of heredity to nucleotide-sequences implied the transition from classical (or Mendelian) genetics to molecular, then – at least in principle – genes can be reduced to physics and chemistry. Thus, as genes contain the information for the structure and function of all live entities, biology could be reduced to physics and chemistry. As in the former case, the question of the reduction of biology triggered once again the epistemological discussion about the unification of natural science by physics (see Ernst Mayr, 1982; Sarkar, 1989). As already outlined in our overview of Part II, the papers in this section of our volume tackle precisely the issue of reductionism and contribute to a new, challenging approach to the distinction between theoretical and observational terms.

Genes and proteins need each other: instructions cannot be carried out without agents, but blind agents cannot fulfill intelligent tasks either. However, if there was a natural start and evolution of life, then one of two macromolecules should have been the first, and afterwards they would have occurred together for performing a joint work. This was the view, at least, in the early stages of the development of molecular biology. According to this view the following question arises: Which came first? In fact, the two hypotheses lead to two research programs on the origin of life. In this respect, genetics inspired a scientific idea which had a tremendous influence on our contemporary culture; an idea developed in 1932 by the British evolutionary biologist John Burdon Sanderson Haldane (1892–1964): because of their replication properties, genes were the entities that must have been present at the beginning of life. More exactly, what this thought suggests is that polynucleotides constituted the primeval molecules and then they associated with polyaminoacids in order to optimize their own replication. However, how was it possible for naked genetic substances to progress up to the point of codifying the structure and function of organisms whose complexity is practically unimaginable? Would the process that neo-Darwinian biologists propose for organic evolution, that is, small random changes under the action of natural selection (cf. Küppers, 1989), have been sufficient?

If genes – understood as replicative polynucleotides – were the true entities from which life originated and also contain the blueprints for all organisms and even viruses, then the following philosophical question should also be put: are the genes, the immortal replicators, the units of selection in the evolutionary process? Are they the means through which organisms replicate themselves or are organisms the means of their own replication? This is the idea now known as *The Selfish Gene Theory*, which has its origin and first developments in the experimental work and speculations 1971 of the American microbiologist Sol Spiegelman (1914–1983), though it is usually attributed to Richard Dawkins by virtue of his famous book *The Selfish Gene* (1976).

Nowadays, in the most recent state of research, the genome (the whole hereditary information of an organism that is encoded in the DNA) may be conceived more as a library than as the initiator and conductor of organic development. Indeed, those who deciphered the genetic code tended to think that it is the *gene* that initiates and conduces the organic development. The genome predetermined the organism: it contained all the information, and all the power. The genome was there from the beginning. The nucleus of the first cell after a sperm and an ovocyte had merged contained the whole person it was going to become. But given what is now known of the properties of stem cells, and of epigenetic (environmental) influences throughout the development of embryos, we rather come to the conclusion that this initial approach might have been too one-sided.

Saying that genes are only one piece of the jigsaw does not render biological research less *analytic* (in the Cartesian sense of the term). Biology has always been balanced between two trends, the *analytic* and the *holistic*. One of the main issues underlying the discussion in our book is the dynamic relation between the holistic (i.e. anti-reductionist) and the analytic trend. Furthermore, the holistic view is re-emerging, in spite of biology going molecular.

The beginnings of the 70s witnessed the arrival of genetic biotechnology. In this sense, the first step was the discovery of the restriction-enzymes that are used by bacteria to defend themselves from viruses or bacteriophages. Because these enzymes cut the enemy DNA into its specific bases, then they could be employed to sequence DNA segments, and for this reason they were denominated *biological cutters*. To this technique of recognition was added that of cloning, by means of which many copies of a given DNA segment could be obtained quickly, in such a way that it could be analyzed exactly in its nucleotide-sequence. Both tools, that is biological cutters and cloning, allowed, for the first time, the DNA and RNA of organisms to be read directly and therefore the reading of genetic messages. These achievements – for which the American physicist Walter Gilbert (1932–) and the British biochemist Frederick Sanger (1918–) were honoured with the Nobel Prize 1980 in Chemistry, opened a wide spectrum of practical applications, among them the possibility of identifying individuals (whether persons or other organisms) and of determining degrees of kinship between them. The same technique also makes it possible to read the entire genome of viruses, such as those of the immunodeficiency family. Without question, the completion of human genome sequence a few years ago, the *Human Genome Project*, has constituted the most outstanding result and a historical achievement in the history of science.

With regard to medical sciences, the technique of DNA reading permitted, in the early 80s, first genetic tests, which were directed at detecting deleterious mutations in our genome and in the genome of other species. Thus, it became possible not only to know the roots of many genetic – mostly hereditary – disorders, but also to evaluate the risk that we have of developing illnesses related to genetic factors, such as the diverse kinds of cancer. In order to ponder the immense importance of these tools for health sciences, it should be noted that whereas 20 years ago tests for gene-based disorders could be counted on the fingers of one hand, today we have more than 900 and this number is increasing.

However, genetic biotechnology is much more than a set of tools for knowing whether genes and genomes are normal or abnormal. It is also an instrument for the modification of living beings or the substance they produce. The discovery of the techniques of recombinant DNA, by means of which it became possible to compose specific DNA sequences, and those of introducing DNA sequences in cells in order for these to fabricate what the genetic messages say, have inaugurated a new, marvellous era. They provide us with insulin and other vital substances for medical treatment, they have transformed properties of some vegetables and animals according to human necessities, and finally, we know that they are the keys to cure or alleviate genetic disease.

For thorough historical further reading let us mention the contributions by R. Olby (1974, 1985) and E. A. Carlson (1966) which constitute outstanding historical guides on the development of genetics, whereas that by M. Morange (1998) *A History of Molecular Biology*. Harvard University Press, Cambridge and by J. Watson and alii (2003) offer a detailed view of this discipline from our contemporary molecular approach. More literature can be selected from the references below.

REFERENCES

Avery O, MacLeod C, MacArty C (1944) Studies on the chemical nature of the substance inducing transformation of pneumococcal types. J Exptl Med 79:137–158

Carlson EA (1966) The gene: critical history. Saunders, Philadelphia

Changeux JP et al. (1963) Allosteric proteins and cellular control systems. J Mol Biol 6:306–329

Dawkins R (1976) The selfish gene. Oxford University Press London

Dunn LC (1965) A short history of genetics. McGraw-Hill, New York

Haldane JS (1932) The causes of evolution. Longmans, London

Hershey AD, Chasey M (1952) Independent functions of viral protein and nucleic acid in growth of bacteriophage. J Genet Physiol 36:39–53

Jacob F, Monod J (1961) Genetic regulatory mechanisms in synthesis of proteins. J Mol Biol 3:318–356

Küppers BO (1989) Information and the origin of life. The MIT Press Cambridge, Mass

Mayr E (1982) The growth of biological thought. Harvard University Press Cambridge, Mass

Nirenberg M, Matthaei JH (1961) The dependence of cell-free protein synthesis in *E. coli* upon naturally occurring or synthetic polyribonucleotide. Proc Nat Acad Sci USA 47:1588–1602

Mendel G (1866) Versuche über Pflanzen-Hybriden. Verhandlungen des naturforschenden Vereines in Brünn 4:318–330

Morange M (1998) A history of molecular biology. Harvard University Press Cambridge, Mass

Morgan TH (1928) The theory of the gene. Yale University Press, New Haven

Olby R (1974) The path to the double helix. University of Washington Press, Seattle

Olby R (1985) The origins of mendelism. Chicago University Press Chicago

Ruse M (1973) The philosophy of biology. Huchinson Library, London

Sarkar S (1989) Reductionism and molecular biology: a reappraisal. PhD dissertation, University of Chicago, Department of Biology

Spiegelman S (1971) An approach to the experimental analysis of precellular evolution. Q. Rev. of Biophys 4:213–254

Watson J, Crick F (1953a) A structure for deoxyribose nucleic acid. Nature 171:737–738.

Watson J, Crick F (1953b) Genetical implications of the structure of deoxyribonucleic acid. Nature 171:964–967

Watson JD et al. (2003) Molecular biology of the gene, 5th edn. Benjamin Cummings New York

ACKNOWLEDGEMENTS

This volume has been 2 years in the making and we would like to thank everyone involved for their patience and effort. Each of the papers included here has been through a rigorous referee process and most have undergone at least one revision. While this process has certainly improved the book, it has also come at the cost of a great deal of time and hard work for all concerned. We would like to explicitly express our gratitude for the expertise of Philippe Huneman (IHPST) who provided us with many suggestions to improve the introduction and overall organization of the book, and to the following referees:

Professor John Beatty, University of British Columbia,Vancouver.

Professor Kurt Bayertz, Münster University.

Professor Yolanda Gómez Sánchez, UNED, Madrid.

Professor Bernd-Olaf Küppers, Jena University.

Professor Carlos Ulises Moulines, München University.

Professor Hans-Jörg Rheinberger, Max–Planck-Institut für Wissenschaftsgeschichte, Berlin.

Professor Carlos Romeo-Casabona, University of País Vasco, Bilbao.

Professor Sahotra Sarkar, University of Texas at Austin.

Special thanks to Lina Lanoir, Marc Kirsch (Collège de France) and Charles Wolfe, who respectively recorded, typed, worked out and translated the interview launching the book, and to Juan Redmond (Lille3), who lent us his typesetting skills with expertise and generosity.

Anne-Fagot Largeault would like to thank Professor François Jacob, Collège de France, who is an awe-inspiring source of insight far beyond the field he has co-created.

Juan Manuel Torres is thankful to the GULBEKIAN Foundation and the FCT – Fundação para a ciência e a tecnologia for financial support, to Prof. Olga Pombo (Universidade de Lisboa) for her academic and personal assistance during his stay at the Universidade de Lisboa, and to Fylis Barrantes for revising the English of some of his texts.

Finally, Shahid Rahman is grateful to André Laks, director of the project "Langue, Textes, Contextes" at the MSH-Nord-Pas de Calais, Laurence Broze, director of the MSH-Nord-Pas de Calais and Fabienne Blaise and Philip Miller, director and subdirector of the Laboratory UMR 8163 "Savoirs, Textes, Langage" (Lille3) for fostering an efficient and collegial research environment and would like to dedicate this book to Cheryl Lobb de Rahman, Jasmin Renschler and Dr Donya Young.

LIST OF CONTRIBUTORS

Francisco José Ayala Department of Ecology and Evolutionary Biology, University of California, Irvine, USA.

Frederic Bouchard Department of Philosophy, University of Montreal, Montreal, Canada.

Miguel Ángel Capó, Camilo José Cela-Conde and Marcos Nadal Laboratory of Human Systematics, University of the Balearic Islands, Palma de Mallorca, Spain.

Claude Debru Chair of Philosophy of Science, Ecole Normale Supérieure, Paris, France.

Anne Fagot-Largeault Chair of Philosophy of Biological and Medical Sciences, Collège de France, Paris, France.

Jean Gayon Institut d'histoire et de philosophie des sciences et des techniques (CNRS/ENS), Paris, France.

Jean-Yves Goffi Université Pierre Mendès France, Grenoble 2, UFR SH, Grenoble, France.

Jonathan Michael Kaplan Philosophy Department, Oregon State University, Corvallis, USA.

Pablo Lorenzano Universidad Nacional de Quilmes – Consejo Nacional de Investigaciones Científicas y Técnicas, Quilmes, Argentina.

Koichiro Matsuno Department of BioEngineering, Nagaoka University of Technology, Nagaoka, Japan.

Michel Morange Centre Cavaillès, Ecole Normale Supérieur, Paris, France.

Shahid Rahman, Professeur de logique et épistémologie, Philosophie, Université de Lille3, Lille, France.

Carlos M. Romeo-Casabona Professor in Criminal Law, Director, Inter-University Chair BBVA Foundation – Provincial Government of Biscay in Law and the Human Genome, University of Deusto and University of The Basque Country, Bilbao, Spain.

Juan Manuel Torres Centro de Filosofia das Ciências, Universidade de Lisboa, Lisboa, Portugal, Universidad Nacional del Sur, Bahía Blanca, Argentina.

Linda van Speybroeck, Gertrudis van de Vijver and Dani De Waele Department of Philosophy and Moral Science, Ghent University, Ghent, Belgium.

INTERVIEW OF FRANÇOIS JACOB (FJ) BY ANNE FAGOT-LARGEAULT (AFL)

FRANÇOIS JACOB
Institut Pasteur, Paris, September 14th 2004[1]*. Nobel Prize in physiology or medicine 1965 Institut Pasteur. Collège de France*
fjacob@pasteur.fr

Recorded and transcribed by *Lina Lanoir*, revised by *Marc Kirsch*
Translated by *Charles Wolfe*, revised by AFL

1. "THE METHODOLOGICAL AND EPISTEMOLOGICAL QUESTION: THE PERSPECTIVE OF MOLECULAR BIOLOGY"

AFL – One hears a lot of talk nowadays about the "geneticization" of medicine and culture. The question lies at the center of debates among sociologists, philosophers and historians in journals like the American *Journal of Medicine and Philosophy* or the European *Medicine, Health Care and Philosophy*. One prominent claim is that the "molecular" trend in medicine, since the middle of the 20th century, is dehumanizing, as it promotes a kind of analytical "pointillism" at the expense of a more holistic approach to the treatment of patients. One hears of the threat of a tyranny of the "genetically correct" (the hunt for bad genes). It is suggested that the use, in human reproduction, of techniques borrowed from genetic engineering gives rise to dangerous tinkering. These problems will serve as the background for our discussion. First, I would like to ask you a methodological and epistemological question: what is the particular perspective, practically speaking, of working molecular biology?

[1] As already mentioned in the introduction, we recall the reader that after the reading of the written version of the original source in French, some paragraphs have been revised. In order to leave the original interview intact we implemented the revisions in the English version only.

You speak a great deal of molecular analysis. In an article with the evocative title "L'embryologie devenue moléculaire"[2] ("Embryology having become molecular"), you state that developmental biology has moved from being descriptive to being analytic. The analytical method was therefore dominant for some time.

"The whole of the living world thus resembles a kind of giant construction set. The same pieces can be taken apart and reassembled in various ways, so as to produce different forms. But basically it is always the same elements which are used."[3] But you also speak frequently of regulation, for instance: "What distinguishes a hen from a fly or a giraffe is not the nature of their components, but the relative proportions and distribution of these components. It is not a matter of structure, but of regulation."[4] This leads me to ask if the molecular standpoint is more attentive to structure or to regulations?

FJ – Both aspects are present. Obviously, the molecular standpoint pertains to the structures of molecules, but at the same time it also pertains to the way in which they are expressed, and to that which modulates their expression: necessarily there has to be a bit of both.

AFL – What was the positive effect of the introduction of this molecular medicine and this analytic standpoint? Conversely, which aspects of it remain incomplete? The first publication was Linus Pauling's "Sickle cell anemia, a molecular disease," a bit earlier than the 1950s. This marked a considerable shift in outlook.

FJ – Yes, and we owe this shift to Pauling.

AFL – People say that physicians were accustomed to consider individuals as a whole and now they focus on the level of the element, which is an anthropological catastrophe.

FJ – That is the anthropo-catastrophe. The process is the same in all technological disciplines. When physicists try to understand how their systems work, they look towards the pure parts, they analyze, take layers apart, and move towards more or less elementary levels. Molecular biology and molecular medicine have the same attitude: they focus on the elements composing the individual, in order to see if certain pieces should be changed, and nothing else. In contrast to this, there is always the opposing attitude, that of the holists, who only consider the whole and are totally uninterested in details and the ways pieces are taken apart. But the analytic attitude is not unreasonable. In any system, every time one thinks one is dealing with a kind of mechanics composed of interacting pieces, and there is a problem somewhere, one tries to find the defective piece and change or correct it.

---

[2] *Médecine/sciences* (1989) 5:6–7.

[3] Jacob J (1997) *La souris, la mouche et l'homme* Odile Jacob, Paris, 130.

[4] Jacob F (2000) Génétique: vraies questions et faux débats, *Le Débat*, 216–224, here, 222.

AFL – Are your models of living beings primarily mechanical?

FJ – Yes, I guess they are Cartesian.

AFL – Yet you say that the study of the regulations is equally important.

FJ – Of course, because in the final analysis, it is regulations which determine if the system is working or not. The regulation elements play a key role. For instance, one has a protein on the one hand, and so-called "regulating" genes on the other hand, which determine if the protein will be expressed or not, in which conditions, in terms of specific interactions with the environments which determine expression or non-expression. The gene has a decisive role: depending on the environment, the actions and interactions it has with other genes, it will push in one direction rather than another, in a fairly mechanical fashion.

AFL – So the standpoint of regulation is not opposed to the standpoint of the element, then?

FJ – No. I believe that they are complementary.

AFL – To tell the truth, I remember that at one time, this analytic method was viewed as something very troubling: to take apart an individual into her parts, and then into so many molecules...

FJ – There have always been two schools: those who want to decompose and examine the elements, and those who are only interested in the whole. I remember debates during my lectures at the Collège de France, when some audience members would protest that "A man is a man. He cannot be cut into pieces in this way." Clearly, there are two parties, two manners of proceeding, which are strongly opposed to one another.

AFL – But if you locate yourself on the analytical side, aren't you more or less against the holists?

FJ – I have nothing against them. In this kind of opposition between two parties, each standpoint has its merits. Both must be taken into account.

AFL – But you understand regulation as something mechanical?

FJ – Everything depends on what kind of regulation we are discussing. In the case of biological regulation, which enables a system to function, which leads it to produce certain substances in greater or smaller quantities, etc., the system appears to be completely mechanical. In any case, things can be interpreted this way. There are probably also quite different explanations.

AFL – So the scientific approach is primarily a mechanical approach?

FJ – Not necessarily, not in all cases. If you put it that way, the claim is too strong. But I do think that the analytic approach is mechanical. One starts with the elements and reassembles them, one reconstructs something based on the elements. In fact, one is trying to put together the pieces of the puzzle, to reconstitute it so that all these elements will function together. That's Descartes.

AFL – You're describing a "bottom-up" method, whereas physicians might be inclined to adopt a "top-down" method. They begin with signs, descriptions of illnesses, and then try and locate the determining factors, the causal factors.

FJ – That is probably true, but ultimately, they too arrive at an analysis: they end up focusing on elements and parts. Their aim is to find the piece that isn't working. To tell the truth, when one focuses on the body, it is rarely considered as a whole. In a relationship, when we are dealing with a person, we treat them as a whole, but it is an abstract holism. When one sees a patient, when one tries to find out why she is ill, generally one tries to understand which system or piece is not working. This is necessarily an analysis.

AFL – Can one analyze regulatory disorders in this way, for example hyperthyroidism?

FJ – Yes, in hyperthyroidism something is produced in excessive quantities. It is a malfunctioning of a regulatory system. I think that it is a good idea at times to take the opposite standpoint, but I also think it is difficult to avoid using analytical models in most situations. We are confronted with complex systems composed of multiple parts, and it is rare that the whole of the system is affected. There is almost always an element of the system or a part of the system which is faulty, and that we try to identify in order to correct it.

AFL – When Claude Bernard invented physiology, he tried to put forth a different vision, with his idea of the internal environment (*milieu intérieur*). This perspective is different from the properly anatomical perspective which had been prevalent up until then. Do you support this anatomical perspective?

FJ – Let's say that I am a Pastorian, a follower of Pasteur.

AFL – But for the Pastorians, the enemy is outside, whereas you ultimately found that the enemy is internal.

FJ – That is true, but I think that Pasteur could just as well have discussed interactions between internal elements. Bacteria do interact with something inside an individual.

AFL – Pasteur himself did not go that far, but how far back do you locate this concern in the Pastorian tradition with internal rather than external enemies?

FJ – External enemies have monopolized the landscape since Pasteur, and roughly until us. It was our group that introduced genetics into the system. But I would like to emphasize this point: in any affection, there is always an external element plus a factor of internal interaction; this is where we overlap with genetics. Even in interactions with proteins, we are already dealing with genetics, since the protein is the expression of a gene. Ultimately, it's always genetics. This is probably what is known as geneticization.

It is a recurring attitude: in all systems, we seek to analyze in order to correct. Physicists look at particles, which for them are the root of all things. For biologists, it ends up being genetic elements, or elements which are derived from genetic structures. At the Institut Pasteur, we took over this idea fairly easily: there are elements, there are external factors and there is something internal that reacts to these external factors. When things are stated in this way, it is difficult to deny it: it is obviously true, in one guise or another. Whether we should speak of genes or not is debatable, and is in part an issue of nomenclature, but there is obviously a schema of this kind: a bacteria appears, and it disturbs something in the individual. The bacteria is not alone in the world, and does not act on the entire organism; thus we end up having to look for an internal factor.

AFL – Pasteur's strategy was to modify circumstances through vaccination. But he didn't know what kind of modifications he was producing. He hadn't had the idea of a cellular or molecular analysis.

FJ – Cells were known in Pasteur's time, people knew that individuals were composed of cells. But cellular properties themselves were not yet really studied, at least not in France.

AFL – Did people at the Institut Pasteur become interested in the molecular effects of vaccinations at the same time as you were interested in these issues, independently?

FJ – It was the group to which I belonged that first adopted a molecular approach, at the Institut Pasteur. Compared to the Americans and those who were working under Pauling's influence, we were certainly more holistic in France. I think we are still the last to be holistic.

AFL – Should one conclude from this that the analytic standpoint is indispensable?

FJ – I think that it is both indispensable and insufficient. One cannot do without it: this is true in physics, and more or less in all disciplines. When we try and understand a system – especially when the situation involves interlocking systems, in which the higher level is comprised of the lower levels –, we need to understand the functioning of each level in order to grasp the higher level, and make sense of the whole. I don't know if this is equally the case in the human sciences.

AFL – With respect to the human sciences, what do you think of the principle of methodological individualism, that is, the idea that we should assume the whole is made of individuals, so that in order to study the whole, we have to focus on the individual? This is the approach of an entire sociological school. For them, methodological individualism amounts to the claim that a society is composed of the actions of individuals. It is by studying individual decisions that one can understand social movements.

FJ – To tell you the truth, I am not familiar enough with the human sciences to answer this. But on this topic of the relation between the individual and the group, there is a question that has always fascinated me: that of the swarms of gnats which follow you as you walk, in a cloud, without actually being concerned with you. How can a system of this sort work? It seems to me that it is not enough to return to the individual, it is more complicated. Similarly, in the human sciences, I suspect that there are global effects which are irreducible.

AFL – And when you speak of embedding [*emboîtement*], I gather that at each level there are specific laws or characteristics, which are irreducible to a lower level. Consequently, your analytical standpoint is not reductionist.

FJ – No. Reductionism is ideological.

AFL – What is the source of your notion of the embedding of modules? The idea plays an important role at the end of *La logique du vivant* (*The Logic of Life*),[5] but it is not very present at the beginning of the book: one gets the impression, as one reads the book, that you introduce linguistic metaphors, and the notion of a program, and then you introduce this notion of modularity which is probably more important. Some of your metaphors are very evocative, such as the box made of boxes, like a nest of dolls, and so is your notion of integration which is close to that of regulation.

FJ – Indeed, for there to be integration there must be regulation.

AFL – And in order to move to the higher level, there must be a regulation which is not reducible to the elements. It is probably at the level of relations that this shift becomes possible, but are relations themselves analyzable? In order to account for this system of embedded modules, one must assume that at the level of the whole, relations exist which are not reducible to what we find at the level of the elements. The interaction between elements at one level creates an environment which gives rise to new structures, which only appear at the higher level. This is what happens in the swarm of gnats: no individual gnat is a basis for understanding the form of the swarm. The question the philosopher will have when reading your work is, how does this integration come about?

FJ – In fact, in the swarm, the form is not constant: the swarm is modulated and in constant agitation throughout its flight. It's extraordinary.

AFL – But isn't an organism as a whole just as extraordinary a phenomenon? Its cohesion and its permanence do not reside in any of its parts. The image of a mechanical assemblage of parts doesn't yield the structure of the whole, and doesn't explain it: a biological assemblage is something very specific. In the case of the embedded modules, neither the structure of the module, nor that of the organism or the cell – which have a certain permanence – can be found at the level of the elements. Indeed, the elements are constantly being replaced. We replace our cells and maintain more or less the same form.

Claude Bernard and Bichat emphasized this point. The functioning of the whole is not something which reduces to a lower level. All elements can be replaced, but the structure remains, overall functioning continues. Metabolism is both a kind of perpetual motion and an invariant condition. It maintains the form of the organism even if its elements are constantly

[5] Jacob F (1974) The logic of life. A history of heredity, trans. Spillmann B Pantheon Books, New York.

changing: they live, die, are replaced, etc., so long as the organism takes in nourishment, eliminates waste, breathes, and maintains various exchanges with the environment.

How do you understand this permanence, which is something evolving, since organisms are always aging and changing?

FJ – The issue here is the existence of a plan: at all levels of living beings and the organism, we notice that there is a plan, but obviously not an architect.

AFL – This is one of the oldest problems of biology?

FJ – And most likely it will remain such for quite a while.

AFL – You used to employ the metaphor of a program, to express this. Nowadays you seem to favor the metaphor of a cloud or a swarm, instead...

FJ – Yes, but it's just as hard to explain, for the time being. In the case of cells, given the height and overall form of an individual composed of cells, the whole is something more robust, more coherent than a swarm of gnats. The cells are in direct interaction, whereas the gnats have to send signals – whether hormones, super-hormones or an entirely different kind of signal, such as flight rhythms. Cells communicate in a variety of ways, by contiguity or at a distance.

AFL – How can such phenomena be explained?

FJ – In fact, in the case of the cells, there are a greater number of possible interactions. It is probably simpler to start by looking at the interactions between gnats and the coherence of their flight pattern.
The goal is to develop some models, whether mathematical or biochemical.

2. "DO YOU WISH TO KNOW WHEN AND HOW YOU WILL DIE?"

AFL – My second question follows from the surprise I felt when reading your book *La souris, la mouche et l'homme*, in particular.

You ask the following question: "Do you wish to know when and how you will die?" You initially seem to argue that it is better not to know, and note that people who know they are carriers of a genetic factor predisposing them to certain illnesses, bear a heavy burden – as in familial predispositions to certain types of cancer. The more we know, you suggest, the more these people will be viewed as pre-patients, which will make their lives consequently more difficult. Nevertheless, three pages later, you state that "it is not knowledge which is dangerous, but ignorance." In Chapter 6, you write that "medicine must move forward; our medicine is founded on genetics, this is a good thing, and ignorance is more dangerous than knowledge. The more medicine knows, the better it can produce diagnoses, and the more people will know about which factors of predisposition they are carriers of." Is it really better to know, ultimately, even if it's painful?

FJ – It seems to me that what is truly difficult is knowing when one will die. But obviously, predictions in this area are difficult to make.

AFL – Indeed. But if we grant that someone can know if she has a predisposition factor, for example, to diabetes or manic depressive disorders, does this ruin her life? Is this kind of knowledge unbearable for people, in personal terms, and independently of consequences such as insurance? It is clear, besides, that knowledge is sometimes an advantage: to know that one is predisposed to diabetes allows one to take precautions. In many cases, it is probably more of a hope than of a concrete reality. For instance, we hope that stem-cell research will be able to address a variety of pathologies. But often, it is genuinely useful to know that one has a predisposition to a particular condition. One can avoid transmitting the illness to one's children, ask for pre-natal diagnoses, etc. This is quite important. Isn't there something disconcerting about the idea that we shouldn't know?

On my part, I have noticed that doctors customarily used this argument – "people don't want to know" – in order to not tell patients the truth. This is less and less true, however: as they gradually inform patients, doctors realize that patients actually want to know. Do you think there is a right not to know?

FJ – I think that one can't generalize: reactions vary from one individual to the next. I don't know if one can speak of rights here. In contrast, I believe that if one were to tell people at age 20, "you will die on this day, at this time", that would make their life harder.

AFL – Granted, but in real life things are less specific than this. One cannot predict death so precisely, but one can say to a woman, for instance, "you are genetically predisposed to breast cancer". Is that harmful to her, would it ruin her life?

FJ – In a sense, it is. It means she will have to be especially careful. Yet I still think it is better to know this kind of thing.

AFL – American women, when they learn that they are carriers of this kind of genetic predisposition, may have recourse to surgery, that is removal of the breast. In fact, it's not clear that this is enough to ward off the danger. In France, this information is not given as readily to women. Certain cancer research centers hesitate to offer this diagnosis. It appears that in some places, they have been reluctant to make the diagnosis even in women about whom there is a reason to believe they might be so predisposed. Now, in the latter cases, knowledge – in the form of a genetic diagnosis – is positive and useful. I am quite opposed to the notion that women should be protected from knowing.

FJ – I disagree on one point. It was true some years ago that patients did not always have access to proper information. But no longer now. For example, in the case of prenuptial exam, at a time when no treatment for syphilis was known, doctors who were aware of one engaged person having syphilis were not allowed to reveal the diagnosis to the other person. Today, however, women would have to be informed. Same thing with cancer: information and consent to treatment are the rule now.

AFL – I've heard people declare that, should they have cancer, they would prefer not to know. And what about AIDS? Some people choose not to know whether they are HIV-positive, despite the risk of contaminating others. Is it legitimate to ask not to know, in these cases? If a doctor discovers that someone is HIV-positive, aren't they under an obligation to tell them?

FJ – Of course, in the case of an infectious disease, a contagious disease such as AIDS, it is not the same thing. Other people are being put in danger, which is unacceptable. But this is a very different situation from that of cancer. Note that 50 years ago, before the release of antibiotics, the problem with syphilis was exactly the same as it has been with AIDS or cancer. This is therefore not a new question, but a recurrent one.

3. "GENETICS AND THE ETHICAL DEBATE"

AFL – Let's now turn to matters of moral philosophy. In a major article-interview in the journal *Le débat*, in 2000, you had said that until the 1960s, "biology was still a descriptive discipline... and then suddenly, we noticed, with some horror, that not only could we watch, but we could also intervene and manipulate." Bacteria began to be engineered genetically towards 1975. During the interview, with reference to the Asilomar conference and the beginnings of molecular biology, you thus begin by describing what appears to be a spectacular, and somewhat disturbing turn: from now on, biology is no longer just a science of observation, since we can intervene. Then you comment, which seems quite right, that in fact, this was not the first case of manipulation. Maybe the means and methods were slower, less powerful or precise, but "wheat as we know it, was created by prehistoric cultivators approximately eight thousand years ago, and contains the chromosomes of three different plants. It is already the result of a manipulation, an act of tinkering!"[6] Thus you relativize the degree of novelty of genetic engineering. Man has always transformed the living nature around him.

In addition, in *La souris, la mouche et l'homme*, you observe that if the goal is to eliminate the risk and consequences of a genetic anomaly, selecting (*trier*) is sufficient: "genetic therapy seems pointless here" (191–192). You had made a similar comment to me earlier on, regarding in vitro fertilization and pre-implantation diagnosis (*diagnostic pré-implantatoire*). Then, you had said that "we don't need to engineer defective embryos, we only have to select (*trier*)." You added that it would be ridiculous

[6] Jacob Génétique: vraies questions et faux débats 218.

to pick abnormal embryos and try to correct the anomaly before transferring them into a woman's womb (an inclination that was present in some Christian militant groups at the time), since one only had to check which ones were normal, and transfer the normal ones. At that time, pre-implantation diagnosis was viewed with suspicion, on account of the selection process. After it was practiced for some time, the practice became acceptable.

FL – This took some time, and the Comité national d'éthique resisted it.

AFL – Once the French National Committee for Ethics was confronted with existing practice, and the success that had been achieved, it accepted this position. But there is another case in which it is harder to maintain your position, prior to implantation, as you mention in *La souris, la mouche et l'homme*. You say that if the point is to give the embryo a new property, such as a resistance to AIDS or cancer, then the goal has changed. It is no longer a matter of selecting the good ones as opposed to the bad ones, but rather, one would have embryos lacking a property we want to give them, so they would have to be injected with it, before being implanted. This is something you reject. "We are dealing with the genetic heritage of humanity here. It is no longer a matter of caring for human beings, but of modifying them, shaping them. All biologists seem to agree that this is to be avoided at all costs."[7]

FJ – I think that is quite right.

AFL – Miroslav Radman gave a presentation some time ago at the Royal Society in London, on experiments that have been done on mice to which P53 genes have been added, which protect against cancer. This work earned him a full page in *Libération* this summer, entitled "We will soon have to start asking about transgenic man." We can see how far these ideas can go, for example with a young father, a responsible individual who knows that in his family, people have had a tendency to develop cancers at the age of thirty ... Of course, at best we still know that this can't be transposed to humans directly, there will be a long research phase. But how is this different in principle from what Pasteur did with vaccinations? Aren't vaccinated beings immunologically modified beings?

FJ – Undoubtedly so, but they are not genetically modified. It is not a modification which will be passed on to their descendents. Nevertheless, I do think that in very particular situations, we will probably come to use these kinds of methods, for instance, if we discovered a way to produce resistance to AIDS. I probably did not express myself well on this point, in my book. My position is that it is better to avoid modifying the human genome, except in very particular cases, such as AIDS. But we need to be very careful, or else we will see all kinds of things develop.

---

[7] La souris, la mouche et l'homme, 191–192.

AFL – Should we always assume that people will behave irrationally? Claude Debru, in his book on biotechnology, takes the opposite position. He points out that at the time of the Asilomar conference, people imagined that scientists were out of control, and that restrictions were needed. What happened, ultimately, is that they used yeast to produce insulin: diabetics benefited greatly, and nothing abominable happened. So why should we assume that a modification of the human genome, in particular cases – or more modestly, to be realistic, a usage of gene therapy in high-risk families – would be so threatening?

FJ – One should simply avoid spreading the fantasy of the "theoretical" child, that possesses all qualities. The perfect baby is such a temptation...

AFL – I agree with you, when you speak of a kind of procreatic fury or mania, in which people want to have babies at any cost. But I don't think they are so attached to the perfect baby. They want a baby, any baby, so long as it is their own. Cases have come up, in the United States but also in France, of deaf parents who wanted their child to be deaf. And you've spoken of cases of medically assisted procreation in which sterility is transmitted to the child.

FJ – Yes, this is a scandalous practice, which has been carried out directly on humans, without any animal experimentation.

AFL – It's true that it's scandalous, but these are isolated cases, which do not involve the human genome. And to get back to that issue, you say that we shouldn't touch the human genome – yet it is in constant mutation. It is always changing. Why shouldn't we be allowed to "touch it up" a bit?

FJ – It modifies itself on its own. But I am very suspicious of the "touching up" we might want to add.

AFL – Yet in the case of wheat, great successes were obtained.

FJ – Yes, but how many failures were required to arrive at this success! It is difficult to take such risks, in the case of human beings.

AFL – Is it merely a question of prudence? Do you think it was necessary to have a universal declaration on the human genome, which stipulates that it must not be interfered with?

FJ – No, I don't think the universal declaration was required.

AFL – Moreover, it declares both that the human genome is not to be modified, and that scientific research must not be impeded.

FJ – Everyone knows that these are opposed to one another.

AFL – I would like to know your position on this issue of the respect of the genome, and the question of whether we have the right to intervene at the genetic level, or not. You seem rather skeptical about gene therapy.

FJ – Gene therapy has not had much success so far, aside from the results of the team at the Hôpital Necker. But with the human genome, if we begin to intervene, however carefully at the outset, I don't know how it will end. There is already talk of reproductive cloning, which I find disturbing. At the same time, I fail to see what is interesting in this method, and I doubt if it could be generalized. It doesn't appear to require universal regulations. On another note, I have nothing against therapeutic cloning.

AFL – But you have reservations about modifications of the genome?

FJ – I am quite reserved except in particular cases, for instance, if it gave us a means to resist a terrible disease such as AIDS.

AFL – Miroslav Radman speaks of a resistance to tuberculosis, malaria, cancer and AIDS.

FJ – Yes, for instance. But the mechanisms of these resistances are complex, as they are not due to monogenic factors. And polygenic factors are very difficult to control.

AFL – That is undoubtedly the reason why Radman doesn't give any other example than the P53 gene, which appears to be a proven factor in mice. There seems to be no equivalent for AIDS.

FJ – But there are cases of people who are AIDS-resistant. Hence there probably are resistant genes.

AFL – So, since vaccination against AIDS doesn't seem to be very promising, maybe this is the best path to follow? The risk then is that the virus will mutate in a way that enables it to avoid the obstacle.

FJ – That is its usual behavior.

AFL – In any case, you don't think it is wrong in principle to abstain from intervening.

FJ – I think biologists are in general agreement that it is better not to intervene. There are cases in which one might be led to intervene, but they are rare.

AFL – Aren't we in a situation akin to the beginning of genetic engineering, when scientists started to modify the genome of bacteria. People were afraid of new developments. There were large-scale warnings, people feared ecological catastrophes, and the like. In fact, nothing really bad happened. Actually, to intervene at the level of the genome is highly complicated, it can't be done just anywhere, it requires teams, labs, scientific and ethical oversight, etc. There are guarantees.

But I would like you to be more specific on another point. You assert quite strongly that "there is no equality in biology."

FJ – Yes, because equality is a cultural concept. There can only be equality between different people. If everyone was identical, we wouldn't need equality. It is only because we are different that we require equality. Hence it is a cultural rather than a biological concept. One cannot say that two molecules are "equal," that would be meaningless. In biology, we are all different.

AFL – Above all, some people have biological handicaps, particularly for genetic reasons. By virtue of a principle of equality, you do hold that patients must be cared for, and that the situation of those who are predisposed to certain pathologies should be improved. You accept that this is undertaken in the form of genetic intervention, e.g., that some day one might try to implant a P53 gene (or its equivalent) to people who run the risk of developing an early cancer, in order to increase their life expectancy. The difficulty here is that these are technologies of developed countries, which run the risk of reinforcing other inequalities. What should one do in this case: develop such techniques in the name of a principle of equality, or give them up, so as not to create inequalities?

FJ – This is an important question. I think in fact, the two arguments are quite different. And a great deal of social issues are involved here.

AFL – Out of egalitarianism, should we choose to forgo curing our defects because we cannot provide the same care for all? This goes back to the kinds of debates that occurred over eugenics, the dream of superior races, etc. Imagine a situation today in which developed countries could offer their citizens a gene that would protect them against cancer. It is clear that the same offer could not be made to all human beings in India or Africa, for example. There will thus be two types of populations.

FJ – This reasoning applies in many cases.

AFL – So should we then say, as in the case of AIDS, that initially, new treatments worsened the level of inequalities, but that this is a transitional period, such that ultimately, everyone will have access to such treatments? How do you see this?

FJ – Ultimately, the real question is what people would think, if they were consulted.

AFL – Lastly, what would you say to the critiques of the "geneticization" of medicine, and to those who argue that the analytical attitude no longer treats the individual as a whole, but merely looks for the defective part, in a hunt for bad genes?

FJ – I would answer that there is no point in expecting this strategy to deliver everything, but that it must be used. I don't think we can do without this approach, and I don't see why we should do without it. But it cannot be the sole approach. We must also seek to grasp things as a whole: genetics is not the only way. One cannot dispense with "geneticization", but it is only a tool.

AFL – And what about medically assisted procreation, and the direct application of genetic engineering to the embryo?

FJ – I was always very reserved as to medically assisted procreation, and the assertion of the right to a child. What is legitimate is the right to health, for people who are sick. We must do what we can to improve their condition. As for the right to a child, or to medically assisted procreation, I am more reserved. In fact, once one can apply such techniques, there is an additional risk, of other techniques being applied to children. Where will this stop? Moreover, in this particular area, techniques are often applied without any research protocol, without any prior animal research.

AFL – Yes, without a protocol or a handbook of experiments. Do you think this is dangerous? That these practices should be forbidden, or controlled?

FJ – I think there will have to be a bit of order here, yes.

AFL – Perhaps it might be appropriate for these techniques not to be reimbursed?

FJ – That would be a more brutal, but also, probably, a more efficient way to do it.

AFL – In any case, the question remains of the right to a child. Do you think the current doctrine is not right?

FJ – I think we should do whatever is possible so that women may have children when they wish, but we cannot do so at any cost.

AFL – Is this tantamount to a kind of respect for nature, at bottom?

FJ – Perhaps. I was quite struck by some work I saw which showed that most lesions occurring during embryo development led to spontaneous abortions: this is an amazing natural system to select what works. Of course, we know that there are also normal fetuses which are aborted in this manner, but in the vast majority of cases, they are abnormal fetuses. This is a fairly enigmatic process. What is it that tells the whole of the system that something is wrong?

AFL – Granted, this is an argument to respect spontaneous abortions and not try to force the continuation of pregnancy when there could be such an abortion. But it is not a reason to forgo trying to have children in vitro.

FJ – The goal is to satisfy people's aspirations, but there are undoubtedly limits that should be respected. Consider the case of premature births.

AFL – One can also ask, just how harmless in vitro procreation is: it may in fact lead to malformations or anomalies in the course of development. So far, studies in this area haven't yielded any decisive results. But in the early days of medically assisted procreation, people avoided such studies quite conspicuously.

In any case, you think there are natural limits – premature birth, spontaneous abortion, etc. – and that nature regulates itself. Hence we should intervene as little as possible. Yet medicine is interventionist, and it corrects nature.

FJ – It is true that it has had some real successes.

AFL – So how should we behave towards what we call nature? Should we respect it or acknowledge that in reality, nature makes mistakes which we can correct?

FJ – Yes, I believe that nature makes mistakes. To correct these mistakes is undoubtedly one of the foundations of medicine. As to the question of whether there is a rule to be followed, which would indicate when we should intervene and when we shouldn't, that is very difficult. It is more a matter of particular cases, to be decided on a case-by-case basis, as with euthanasia.

AFL – Yes, there as well, at the end of a life, nature makes mistakes. Would you accept the idea of a "wisdom of nature"?

FJ – To a certain extent.

AFL – To a certain extent, because in fact nature improvises a good deal, it has no intrinsic wisdom: simply, overly deficient organisms will not survive...

ENTRETIEN DE FRANÇOIS JACOB (FJ) AVEC ANNE FAGOT-LARGEAULT (AFL)

Institut Pasteur, Paris, 14 septembre 2004
Enregistré, transcrit par *Lina Lanoir*, retravaillé par *Marc Kirsch*

AFL – On parle beaucoup, aujourd'hui, d'une "généticisation" de la médecine et de la culture. La question est au centre d'un débat soulevé par des sociologues, philosophes ou historiens de la médecine, dans des journaux comme le *Journal of Medicine and Philosophy* (américain) et *Medicine, Health Care and Philosophy* (européen). On a prétendu notamment que la médecine, devenant 'moléculaire' au milieu du 20e siècle, se serait déshumanisée, cédant à un pointillisme analytique au détriment d'une prise en charge globale des malades. On parle de la menace d'une tyrannie du "génétiquement correct" (chasse aux mauvais gènes); on dit que l'utilisation en reproduction humaine de techniques issues du génie génétique ("reprogénétique") occasionne des bricolages dangereux. Ces problèmes seront la toile de fond de notre entretien.

La première question que je souhaite vous poser est une question de méthodologie et d'épistémologie: qu'est-ce que le point de vue de l'analyse moléculaire a apporté de positif? Vous parlez beaucoup d'analyse moléculaire. «De descriptive, la biologie du développement est devenue analytique», écrivez-vous dans un article dont le titre est parlant: «L'embryologie devenue moléculaire» (*Médecine/sciences*, 1989, 5:6–7). La méthode analytique a donc dominé pendant un temps. «L'ensemble du monde vivant ressemble ainsi à une sorte de Meccano géant. Les mêmes pièces peuvent être démontées et remontées de façon différente, de manière à produire des formes différentes. Mais à la base, ce sont toujours les mêmes éléments qui sont utilisés» (François Jacob, *La souris, la mouche et l'homme*, Paris: Odile Jacob, 1997, 130.). Mais vous parlez aussi beaucoup de régulation. Vous écrivez notamment: «Ce qui distingue une poule d'une mouche ou d'une girafe, ce n'est pas la nature de leurs composants. Ce sont les proportions relatives et la distribution de ces composants. Ce n'est pas une question de structure, mais de régulation» (*Le Débat*, 2000, 222). D'où ma question: le point de vue moléculaire est-il plus attentif à la structure ou aux régulations?

FJ – Les deux aspects sont présents. Évidemment, le point de vue moléculaire concerne la structure des molécules mais en même temps il concerne la façon dont elles sont exprimées et ce qui en module l'expression: forcément c'est un peu les deux.

AFL – Qu'est-ce que l'introduction de cette médecine moléculaire ou de cette attitude analytique a apporté de positif et quels sont les éléments qui la rendent incomplète? Il y a eu une première publication qui est de Linus Pauling, un peu avant les années 1950, 'Sickle cell anemia, a molecular disease'. C'était un changement de perspective considérable.

FJ – Oui, et on le doit à Pauling.

AFL – On a dit que les médecins avaient l'habitude de considérer les gens dans leur entier et maintenant ils vont à l'élément et c'est une catastrophe anthropologique.

FJ – Ça c'est l'anthropo-catastrophe. Le processus est le même dans toutes les techniques. Quand les physiciens essaient de comprendre comment fonctionnent leurs systèmes, ils vont chercher les parties pures, ils analysent, ils décortiquent, ils vont aux choses plus ou moins élémentaires. La biologie moléculaire et la médecine moléculaire ont la même attitude: elles s'efforcent de regarder les éléments qui composent l'individu, de voir s'il faut changer certaines pièces et rien d'autre. Face à eux, il y a toujours l'attitude inverse, celle des holistes, qui ne regardent que le tout et se désintéressent totalement des détails et des façons de décortiquer les morceaux. Mais tout de même, l'attitude analytique n'est pas déraisonnable. Dans tout système, chaque fois que l'on estime avoir affaire à une sorte de mécanique composée de pièces qui interagissent les unes avec les autres, et qu'il y a un défaut quelque part, on essaie d'atteindre la pièce défectueuse et de la changer ou de la corriger.

AFL – Vos modèles de vivants sont principalement mécaniques?

FJ – Oui, cartésiens, sans doute.

AFL – Pourtant, vous dites que l'étude des régulations est importante également.

FJ – Bien sûr, parce qu'en fin de compte, ce sont les régulations qui déterminent si le système fonctionne ou non. Les éléments de régulation jouent un rôle primordial. On a par exemple d'une part une protéine et d'autre part des gènes dits régulateurs, qui décident si la protéine sera exprimée ou non, dans quelles conditions, en fonction de quelles interactions avec les milieux qui déterminent l'expression ou la non-expression. Le gène a un rôle décisif: en fonction du milieu, des actions et des interactions qu'il a avec d'autres gènes, il va pousser dans un sens ou dans un autre, de façon assez mécanique.

AFL – Le point de vue de la régulation n'est donc pas opposé au point de vue de l'élément.

FJ – Non. Je crois que c'est complémentaire.

AFL – À vrai dire, je me souviens qu'à une époque, on trouvait cette méthode analytique très inquiétante: décortiquer un individu dans ses parties, puis dans ses molécules...

FJ – Il y a toujours eu deux écoles: ceux qui veulent décomposer et examiner les éléments, et ceux qui ne s'intéressent qu'au tout. Je me souviens de discussions, lors de mes cours au Collège de France, où certains auditeurs protestaient: «L'homme c'est l'homme. On ne peut pas le couper en morceaux de cette façon.» Il y a clairement deux parties, deux façons de faire, très opposées l'une à l'autre.

AFL – Mais si vous vous placez du côté analytique, vous êtes plutôt contre les holistes?

FJ – Je n'ai rien contre eux. Dans ce genre d'opposition entre deux parties, chacun des points de vue a ses mérites. Il faut considérer les deux.

AFL – Mais vous concevez la régulation comme quelque chose de mécanique?

FJ – Tout dépend de quelle régulation on parle. La régulation biologique, celle qui fait fonctionner un système, qui le conduit à fabriquer certaines substances en plus ou moins grande quantité, etc.: dans cette régulation, en effet, il y a apparemment un système complètement mécanique. En tout cas on peut interpréter les choses de cette façon. Il y a probablement aussi des explications totalement différentes.

AFL – L'approche scientifique est plutôt une approche mécanique?

FJ – Pas forcément, pas dans tous les cas. Sous cette forme, c'est une affirmation trop radicale. Mais je pense en effet que l'approche analytique est mécanique. On part des éléments et on les réassemble, on reconstruit à partir d'eux. En fait, on cherche à assembler les morceaux du puzzle pour le reconstituer et faire fonctionner ensemble tous ces éléments. C'est cela, Descartes.

AFL – Vous décrivez une méthode " de bas en haut ", tandis que les médecins auraient peut-être tendance à aller plutôt de haut en bas. Les médecins partent des signes, de la description de la maladie et ils essaient ensuite de trouver les facteurs déterminants, les facteurs causals.

FJ – C'est probable, mais en fin de compte, ils en viennent aussi à une analyse: ils vont s'intéresser aux éléments et aux parties. Leur but est de trouver la pièce qui ne marche pas. À vrai dire, le corps, quand on s'occupe de lui, est rarement considéré comme un tout. Dans la relation, quand on a affaire à une personne, on la considère comme un tout, mais c'est un holisme abstrait. Quand on voit un patient, qu'on cherche pourquoi il est malade, en général on cherche à comprendre quel est le système ou la pièce qui ne marche pas. Forcément, on analyse.

AFL – Peut-on analyser de la même façon les maladies des régulations, comme par exemple, l'hyperthyroïdie?

FJ – Oui, dans l'hyperthyroïdie, un produit est fabriqué en quantité excessive. C'est un dysfonctionnement d'un système régulateur. Je pense qu'il est bon de temps en temps d'adopter l'autre point de vue, mais il me semble aussi qu'il est difficile de ne pas utiliser les méthodes analytiques dans la plupart des situations. Nous sommes confrontés à des systèmes complexes avec des pièces multiples, et il est rare que ce soit l'ensemble du système qui soit atteint. Il y a presque toujours un élément du système ou une pièce du système qui est défaillante et qu'on s'efforce d'identifier pour la corriger.

AFL – Claude Bernard, en inventant la physiologie, avait cherché à promouvoir une autre conception, avec son idée du milieu intérieur, etc. C'était une perspective différente de la perspective proprement anatomique qui prévalait. Vous défendez cette perspective anatomique?

FJ – Disons que je suis pastorien.

AFL – Mais chez les pastoriens l'ennemi est extérieur: or, vous en êtes venu à trouver que l'ennemi est à l'intérieur.

FJ – C'est vrai, mais je pense que Pasteur aurait pu parler aussi des interactions avec des éléments internes. Les bactéries interagissent bien avec quelque chose à l'intérieur d'un individu.

AFL – Pasteur n'a pas poussé jusque-là, mais à quand faites-vous remonter, dans la tradition pastorienne, ce souci pour les ennemis intérieurs plutôt que pour les ennemis extérieurs?

FJ – Les ennemis extérieurs ont occupé tout le paysage depuis Pasteur et à peu près jusqu'à nous. C'est notre groupe qui a introduit la génétique dans le système. Mais j'insiste sur ce point: dans les affections, il y a toujours – c'est là qu'on rejoint la génétique – un élément extérieur plus un facteur d'interaction interne. Même dans les interactions avec des protéines, on est déjà dans la génétique: la protéine est l'expression d'un gène. En fin de compte, c'est toujours de la génétique. C'est sans doute cela qu'on appelle la généticisation.

C'est une attitude récurrente: dans tous les systèmes, on cherche à analyser pour corriger. Les physiciens vont aux particules, qui sont pour eux la racine des choses. Chez les biologistes, ce sont finalement les éléments génétiques ou dérivés des structures génétiques. À l'Institut Pasteur, on a adopté assez facilement cette idée: il y a des éléments, il y a des facteurs externes et quelque chose d'interne qui réagit à ces facteurs externes. Lorsqu'on énonce les choses de cette manière, il est difficile d'y résister: c'est forcément vrai, sous une forme ou sous une autre. Qu'il faille ou non parler de gène, c'est discutable et c'est en partie une question de nomenclature, mais on a forcément un schéma de ce genre: une bactérie arrive, elle dérange quelque chose dans l'individu. La bactérie n'est pas seule au monde, et elle n'agit pas sur l'organisme entier: forcément, on en vient à rechercher un facteur interne.

AFL – La stratégie de Pasteur était de modifier le terrain par la vaccination. Mais il ne se doutait pas des modifications qu'il produisait. Il n'avait pas eu l'idée d'une analyse cellulaire ou moléculaire.

FJ – On connaissait les cellules du temps de Pasteur, on savait que les individus étaient faits de cellules. Mais on n'étudiait pas encore beaucoup les propriétés cellulaires, du moins en France.

AFL – Est-ce que, à l'institut Pasteur, des gens se sont mis à s'intéresser aux effets moléculaires des vaccinations en même temps que vous vous intéressiez au terrain de façon indépendante?

FJ – C'est le groupe auquel j'appartenais qui a lancé les approches moléculaires, à l'Institut Pasteur. Par rapport aux américains et aux courants inspirés de Pauling, nous étions certainement plus holistes, en France. Je crois que nous sommes toujours les derniers holistes.

AFL – Faut-il en conclure que le point de vue analytique est indispensable?

FJ – Je crois qu'il est à la fois indispensable et insuffisant. On ne peut pas s'en passer: c'est vrai en physique, c'est assez vrai dans toutes les disciplines. Quand on essaie de comprendre un système – surtout quand il s'agit de systèmes qui s'emboîtent, où le niveau supérieur est constitué de niveaux inférieurs –, on a intérêt à comprendre le fonctionnement de chaque niveau pour appréhender le niveau supérieur et comprendre l'ensemble. Je ne sais pas si c'est aussi vrai dans les sciences humaines.

AFL – Que pensez-vous, dans ce domaine des sciences humaines, du principe de l'individualisme méthodologique: l'idée selon laquelle il faut supposer que le tout est composé d'individus et que pour étudier le tout, il faut aller à l'individu. C'est la ligne adoptée par toute une école de sociologues. Pour eux, l'individualisme méthodologique consiste à dire qu'une société est faite des actions des individus. C'est en étudiant les décisions des individus qu'on comprend les mouvements sociaux.

FJ – À vrai dire, je connais trop mal les sciences humaines pour répondre. Mais sur ce sujet du rapport de l'individu au groupe, il y a une question qui m'a toujours intrigué, c'est celle des essaims de moucherons, ces nuages que vous voyez se déplacer avec vous quand vous marchez, sans que les moucherons se préoccupent de vous. Comment un système de ce genre peut-il fonctionner? Il me semble qu'il ne suffit pas de retourner à l'individu, c'est plus compliqué. De la même manière, dans les sciences humaines, je soupçonne qu'il y a des effets globaux qui ne sont pas réductibles.

AFL – Et quand vous parlez d'emboîtements, je comprends qu'à chaque niveau il y a des lois propres ou des caractéristiques propres, qui ne sont pas réductibles. Par conséquent, votre point de vue analytique n'est pas réductionniste.

FJ – Non. Le réductionnisme est idéologique.

AFL – D'où vient votre notion de l'emboîtement des modules? Cette idée joue un rôle très important à la fin de *La logique du vivant* (F. Jacob, Paris: Gallimard, 1970) mais elle est peu présente au début: on a l'impression, au cours du livre, que vous introduisez des métaphores linguistiques et la notion de programme, puis on arrive à cette notion de modularité qui est sans doute plus importante. Il y a des métaphores très parlantes: la boîte faite de boîtes, etc., et cette notion d'intégration qui rejoint la notion de régulation.

FJ – En effet, pour intégrer, il faut réguler.

AFL – Et pour passer au niveau supérieur, il faut une régulation qui ne se réduise pas aux éléments. C'est sans doute dans les relations qu'on trouve ce qui permet ce passage, mais les relations sont-elles analysables? Pour rendre compte de cette modularité par emboîtement, il faut supposer qu'il existe au niveau du tout des relations qui ne sont pas réductibles à ce qu'on trouve dans les éléments. Le fait qu'il y ait interaction entre les éléments d'un niveau crée un environnement producteur de structures nouvelles, qui n'apparaissent qu'au niveau supérieur. C'est ce qui se passe dans le nuage de moucherons: les moucherons individuels ne permettent pas de comprendre la forme du nuage. La question que se pose le philosophe en vous lisant est de comprendre comment se fait cette intégration.

FJ – En réalité, dans le nuage, la forme n'est pas constante: le nuage se module, s'agite à mesure qu'il vole. C'est extraordinaire.

AFL – Mais est-ce qu'un organisme dans son entier n'est pas un phénomène tout aussi extraordinaire? Ce qui fait sa cohésion et sa permanence n'est dans aucune des parties. L'image d'un assemblage mécanique de parties ne donne pas la structure du tout et n'en rend pas bien compte: l'assemblage, en biologie, est très spécifique. Dans le cas de modules emboîtés, la structure du module, comme celle de l'organisme ou de la cellule – qui ont une certaine permanence – ne se trouvent pas dans les éléments. D'ailleurs les éléments sont remplacés en permanence. Nous remplaçons nos cellules et nous gardons à peu près la même forme.

Claude Bernard et Bichat ont souligné ce point. Le fonctionnement du tout n'est pas réductible. On peut remplacer tous les éléments: la structure demeure, le fonctionnement global se poursuit. Le métabolisme est à la fois un mouvement perpétuel et une condition d'invariance. Il maintient la forme de l'organisme même si ses éléments ne cessent de changer: ils vivent, meurent, se remplacent, etc., à mesure que l'organisme se nourrit, élimine, respire, entretient des échanges avec l'environnement.

Comment concevez-vous cette permanence, qui est évolutive puisque les organismes vieillissent, changent?

FJ – Le problème est celui du plan. À tous les niveaux du vivant et de l'organisme, on constate qu'il y a un plan. L'architecte, évidemment, fait défaut.

AFL – C'est un des plus vieux problèmes de la biologie?

FJ – Et qui risque de le rester encore pour quelques temps.

AFL – Pour le représenter, vous aviez la métaphore du programme. Vous semblez préférer aujourd'hui celle du nuage...

FJ – Oui, mais c'est tout aussi difficile à expliquer, pour le moment. Dans le cas des cellules, étant donné la taille, la forme générale d'un individu fait de cellules, on obtient un ensemble plus robuste et plus cohérent qu'un nuage de moucherons.

Les cellules interagissent directement, tandis que les moucherons doivent envoyer des signaux – hormones, superhormones, ou signaux d'une toute autre nature, comme le rythme du vol. Les cellules communiquent de bien des manières, par contigüité ou à distance.

AFL – Comment expliquer ces phénomènes?

FJ – En fait, dans le cas des cellules, il y a plus d'interactions possibles. Il serait sans doute plus aisé de commencer par s'intéresser aux interactions entre les moucherons et à la cohérence de leur vol.

Il s'agit de trouver des modèles, mathématiques ou biochimiques.

AFL – Ma seconde question a été suscitée par l'étonnement que j'ai éprouvé en lisant, en particulier, votre livre *La souris, la mouche et l'homme*.

Vous y posez cette question: «Voulez-vous savoir quand et comment vous allez mourir?». Pendant quelques pages, vous semblez considérer qu'il vaut mieux ne pas le savoir. Vous rappelez que les gens qui savent qu'ils sont porteurs d'un facteur génétique qui les prédispose à certaines maladies portent un lourd fardeau – par exemple dans le cas de prédispositions familiales à certains cancers. Plus on en saura, dites-vous, plus ces personnes seront considérées comme des pré-malades, et plus leur vie sera difficile. Néanmoins, trois pages plus loin, vous affirmez que «ce n'est pas la connaissance qui est dangereuse, c'est l'ignorance». Au chapitre VI, vous écrivez qu' «il faut que la médecine avance, que notre médecine est fondée sur la génétique, que c'est une bonne chose et que l'ignorance est plus dangereuse que la connaissance. Plus la médecine en sait, plus elle est capable de faire des diagnostics, plus les gens vont savoir de quels facteurs de prédispositions ils sont porteurs». Vaut-il mieux savoir en fin de compte, même si c'est pénible?

FJ – Il me semble que ce qui est difficile, c'est de savoir quand on va mourir. Mais à vrai dire, les prédictions sont difficiles, dans ce domaine.

AFL – Oui. Mais en admettant que quelqu'un puisse savoir qu'il a un facteur de prédisposition, par exemple au diabète ou à la maladie maniaco-dépressive, sa vie s'en trouve-t-elle gâchée? Ce type de connaissance est-il insupportable pour les gens, à titre personnel et indépendamment des conséquences éventuelles en matière d'assurances, par exemple? Il est clair, par ailleurs, que la connaissance est parfois un avantage: savoir qu'on a une prédisposition au diabète permet de prendre des précautions. Dans beaucoup de cas, il s'agit sans doute plus d'un espoir que d'une réalité déjà effective. Pour certaines pathologies, par exemple, on attend beaucoup de la recherche sur les cellules souches. Mais souvent, le fait de savoir qu'on a une prédisposition à une maladie est utile. Cela peut permettre d'éviter de transmettre la maladie à ses enfants, de faire un diagnostic prénatal, etc. C'est très important. N'est-ce pas une attitude déconcertante que de ne pas vouloir savoir?

Pour ma part, j'observe que les médecins avaient l'habitude d'utiliser cet argument – «les gens ne veulent pas savoir» – pour ne pas dire la vérité aux patients. C'est de moins en moins le cas: à mesure qu'ils informent les patients, les médecins se rendent compte que les gens veulent savoir, en réalité. Pensez-vous que ne pas vouloir savoir est un droit ?

FJ – Je pense qu'on ne peut pas généraliser: les réactions varient selon les individus. Certains veulent savoir, d'autres non. Je ne sais pas si on peut parler de droit. En revanche, il me semble que si l'on disait aux gens, à vingt ans, vous mourrez tel jour à telle heure, ça leur rendrait la vie plus difficile.

AFL – Sans doute, mais dans la réalité, les choses sont moins précises. On ne peut pas prédire la mort avec précision, mais on peut dire à une femme, par exemple, «vous avez une prédisposition génétique au cancer du sein». Est-ce lui faire du tort, gâcher sa vie?

FJ – En un sens, oui. Cela signifie qu'elle devra faire attention. Pourtant, je pense qu'il vaut mieux savoir ce genre de choses.

AFL – Les femmes américaines qui se savent porteuses de ce gène de prédisposition ont souvent recours à la chirurgie et à l'ablation des seins. Il n'est d'ailleurs pas certain que cela suffise à conjurer le péril. En France, on ne donne pas aussi facilement cette information aux femmes. Certains centres cancérologiques refusent de faire le diagnostic. Il semble que dans certains centres, le diagnostic est refusé même chez des femmes chez lesquelles on soupçonne une prédisposition génétique. Or dans ces cas-là, la connaissance – le diagnostic génétique – est positive et utile. Je suis très opposée à l'attitude adoptée en France sur cette question.

FJ – Je partage tout à fait votre avis.

AFL – Et il y a encore d'autres raisons. Dans le cas du SIDA, on sait qu'il y a des gens qui préfèrent ne pas savoir s'ils sont séropositifs, malgré les risques de contaminer d'autres personnes. Est-il légitime de demander de ne pas savoir, dans ce cas? Si un médecin découvre qu'une personne est séropositive, n'a-t-il pas l'obligation de lui dire?

FJ – Bien sûr, dans le cas d'une maladie infectieuse, contagieuse comme le SIDA, ce n'est pas la même chose. On met les autres en danger: c'est inadmissible. Mais c'est une situation toute différente du cancer.

AFL – J'en viens maintenant aux questions de philosophie morale. Dans un grand article d'entretien dans la revue *Le débat,* en 2000, vous disiez que, jusque dans les années 1960, «la biologie était encore une discipline descriptive ... Et puis, brusquement, on s'est aperçu, avec une demi-horreur, que non seulement on pouvait regarder, mais qu'on pouvait intervenir et manipuler» (in Jacob F., 'Génétique: vraies questions et faux débats', *Le Débat*, 2000, 216–224). On a commencé à manipuler les bactéries vers les années 1975. Au cours de l'entretien, évoquant Asilomar et les débuts de la biologie moléculaire, vous commencez donc par décrire ce qui apparaît comme un tournant spectaculaire et un peu inquiétant: désormais, la biologie cesse d'être une science d'observation, on va pouvoir intervenir. Puis vous remarquez, ce qui me paraît très juste, qu'au fond ce n'était pas la première fois qu'on manipulait. On le faisait plus lentement, sans doute, avec des moyens moins puissants et moins précis, mais «notre blé, tel qu'il a été fabriqué par les agriculteurs de la préhistoire il y a huit mille ans à peu près, contient les chromosomes de trois plantes différentes. Il est déjà le produit d'une manipulation, d'un bricolage !» (ibid., 218). Vous relativisez donc la nouveauté de ces manipulations. Il y a toujours eu transformation de la nature vivante par l'homme.

Par ailleurs, dans *La souris, la mouche et l'homme*, vous observez que s'il s'agit d'écarter le risque d'une anomalie, le tri suffit: «la thérapie génique paraît ici sans objet» (*SMH*, 191–192). Vous m'aviez fait une remarque similaire bien plus tôt, à propos du diagnostic pré-implantatoire. Vous disiez alors: «pour le diagnostic pré-implantatoire, on n'a pas besoin de manipuler, il suffit de trier». Vous ajoutiez qu'il serait ridicule d'injecter des embryons anormaux – c'était une tentation, du côté des chrétiens militants, à cette époque – puisqu'on n'a qu'à vérifier lesquels sont normaux, et choisir ceux-là. À l'époque, effectivement, le diagnostic pré-implantatoire était un tri. On en est venu à le réaliser et finalement, cette pratique est devenue acceptable.

FL – Il a fallu du temps, et le Comité national d'éthique a eu des résistances.

AFL – Une fois confronté à la pratique, mis devant le fait accompli – et devant les succès obtenus –, le Comité national d'éthique français s'est incliné. Mais votre raisonnement est plus difficile à tenir dans un autre cas, qui pourrait aussi se présenter au niveau pré-implantatoire, comme vous le signalez dans *La souris, la mouche et l'homme* (*SMH*),. Vous dites que s'il s'agissait de conférer, à cet embryon, une propriété nouvelle, par exemple une propriété de résistance au SIDA ou de résistance au cancer, on changerait d'objectif. On ne pourrait plus trier les bons des mauvais, on aurait des embryons dépourvus de cette propriété et on voudrait la leur conférer, donc il faudrait la leur injecter avant de les implanter. Et vous refusez cette idée. «On touche au patrimoine génétique de l'humanité. Il s'agit non plus de soigner l'homme, mais de le modifier, de le façonner. Et l'ensemble des biologistes semble d'accord: à éviter à tout prix».(ibid.).

FJ – Je crois que c'est vrai...

AFL – Miroslav Radman a fait, il y a quelque temps, une conférence à la Société Royale de Londres, il est question de travaux réalisés sur des souris auxquelles on ajoute des gènes P53 qui ont un effet protecteur contre le cancer. Cet article lui a valu cet été une page entière dans *Libération*, intitulée: «Il va bien falloir se poser la question de l'homme transgénique». On voit bien la portée que peuvent avoir ces idées, par exemple auprès d'un jeune père de famille responsable qui saurait que dans sa famille les gens font souvent des cancers à trente ans... Bien sûr, dans le meilleur des cas, on sait bien que ce n'est pas transposable à l'homme directement, qu'il y aura nécessairement une longue phase de recherche. Mais en quoi est-ce différent, dans le principe, de ce que faisait Pasteur en vaccinant ? Des êtres vaccinés ne sont-ils pas des êtres immunologiquement modifiés ?

FJ – Sans doute, mais ce n'est pas génétique. Ce n'est pas une modification qui se perpétue dans la descendance. Je pense néanmoins que pour des situations très particulières, on en viendra probablement à ce type de méthodes: si on trouvait une résistance au SIDA par exemple. Je me suis sans doute mal exprimé sur ce point dans mon livre. Ma position, c'est qu'il vaut mieux éviter de modifier le génome humain, sauf dans des cas très précis, dont le SIDA est un bon exemple. Mais il faut être très prudent, sinon, on risque de voir se développer toutes sortes de choses.

AFL – Doit-on toujours présupposer que les gens vont faire n'importe quoi? Claude Debru, dans son livre sur les biotechnologies, prend une position inverse. Il rappelle qu'à l'époque de la conférence d'Asilomar, on a imaginé que les chercheurs allaient faire n'importe quoi et qu'il fallait mettre des garde-fous. Ce qui s'est passé, en fin de compte, c'est qu'on a fait fabriquer de l'insuline par des levures. Et les diabétiques s'en sont mieux portés, mais il ne s'est rien produit d'abominable. Donc pourquoi présupposer une telle menace si on se décidait dans des cas particuliers à modifier le génome humain, ou plus modestement – soyons réalistes – à essayer des thérapies génétiques dans les familles à risques?

FJ – Il faut seulement éviter de répandre le fantasme de l'enfant «théorique», pourvu de toutes les qualités. Le bébé parfait, c'est une telle tentation...

AFL – Je suis de votre avis quand vous parlez de l'acharnement procréatique: les gens veulent avoir des bébés à tout prix. Mais je crois qu'ils ne sont pas si attachés au bébé parfait. Ils veulent un bébé, n'importe quel bébé, pourvu que ce soit le leur. On a rapporté, aux États-Unis, mais aussi en France, le cas de parents sourds demandant qu'on fasse en sorte que leur enfant soit sourd. Et vous parlez de cas de procréation médicalement assistée où l'on transmet la stérilité.

FJ – Oui, c'est une pratique scandaleuse, qui a été mise en œuvre directement sur l'homme, sans essai sur l'animal.

AFL – C'est scandaleux, c'est vrai, mais ce sont des cas isolés. Et le génome humain n'est pas en cause. Et pour en revenir à cette question, vous dites qu'il ne faut pas toucher au génome humain: pourtant, il est en mutation permanente. Il ne cesse de changer. Et nous n'aurions pas le droit de lui faire quelques petites retouches?

FJ – Il se modifie tout seul. Mais je me méfie beaucoup des retouches qu'on pourrait avoir envie de lui apporter.

AFL – Pourtant, dans le cas du blé, on a obtenu de belles réussites.

FJ – Oui, mais combien d'échecs pour obtenir cette réussite. Difficile de prendre des risques, dans le cas de l'homme.

AFL – Est-ce simplement une question de prudence? Pensez-vous qu'il était nécessaire d'édicter une déclaration universelle sur le génome humain, qui stipule qu'on ne doit pas y toucher?

FJ – Non, la déclaration universelle n'était sans doute pas une nécessité.

AFL – Qui plus est, elle dit à la fois qu'il ne faut pas modifier le génome humain et en même temps qu'il faut respecter la liberté de la recherche.

FJ – Tout le monde sait que cela s'oppose.

AFL – J'aimerais connaître votre position sur cette question du respect du génome et sur le fait que nous ayons ou non le droit d'intervenir au niveau génétique. Vous semblez assez sceptique sur les thérapies géniques.

FJ – Les thérapies géniques ont eu peu de succès jusqu'à présent, hormis les résultats obtenus par l'équipe de l'hôpital Necker. Mais avec le génome humain, si l'on commence à intervenir, si prudent que l'on soit au début, je ne sais pas comment cela finira. On parle déjà de clonage reproductif. Ça me paraît inquiétant. En même temps, je ne vois pas vraiment l'intérêt de cette méthode, et je ne crois pas qu'elle puisse se généraliser. Ça ne me paraît pas exiger une loi universelle. Par ailleurs, je n'ai rien contre le clonage thérapeutique.

AFL – Mais vous êtes réticent vis-à-vis des modifications du génome?

FJ – Je suis réticent sauf, pour des cas très particuliers, par exemple, s'il c'était le moyen de résister à une maladie terrible comme le SIDA.

AFL – Radman parle de la résistance à la tuberculose, au paludisme, au cancer et au SIDA.

FJ – Par exemple, oui. Mais, les mécanismes de ces résistances sont complexes: ils ne sont pas dus à des facteurs monogéniques. Et les facteurs polygéniques sont très difficiles à maîtriser.

AFL – C'est sans doute la raison pour laquelle Miroslav Radman ne donne pas d'autre exemple que celui du gène P53: là, on semble tenir un facteur qui a fait ses preuves chez la souris. Il semble qu'on n'ait pas d'équivalent pour le SIDA.

FJ – Mais il y a des cas de personnes résistantes au SIDA. Il y a donc probablement des gènes de résistance.

AFL – Et, puisque la piste de la vaccination ne paraît pas très favorable, c'est peut-être la piste à suivre? Le risque, c'est que le virus mute de telle façon qu'il contourne l'obstacle.

FJ – C'est son comportement habituel.

AFL – En tout cas, en principe, vous n'êtes pas opposé à l'abstention.

FJ – Je crois que les biologistes sont assez d'accord pour éviter d'intervenir. Il y a des cas où l'on pourrait être conduit à le faire, mais ils sont très rares.

AFL – Ne serions-nous pas dans une situation comparable à celle qui prévalait à l'époque des début du génie génétique, lorsqu'on commençait à modifier le génome des bactéries. On avait peur de la nouveauté. Il y a eu de grandes mises en garde, on craignait des catastrophes écologiques, etc. En réalité, il ne s'est rien produit. De fait, manipuler le génome, c'est compliqué, cela ne peut pas se faire n'importe où, cela suppose des équipes, des laboratoires, des contrôles scientifiques, éthiques, etc. Il y a des garanties.

Mais j'aimerais que vous apportiez des précisions sur un autre aspect. Vous affirmez, avec force: «il n'y a pas d'égalité en biologie».

FJ – Oui. C'est parce que l'égalité est un concept culturel. Il n'y a d'égalité qu'entre des gens différents. Si tout le monde était identique, on n'aurait pas besoin d'égalité. C'est simplement parce que nous sommes différents que nous avons besoin d'égalité. C'est donc un concept culturel et non pas biologique. On ne peut pas dire que deux molécules sont «égales», cela ne veut rien dire. En biologie nous sommes tous différents.

AFL – En particulier, il y a des gens qui sont handicapés biologiquement, notamment pour des raisons génétiques. Vous admettez, en vertu d'un principe d'égalité, qu'on soigne les malades et qu'on essaie d'améliorer la situation de ceux qui ont des prédispositions à certaines pathologies. Vous admettez qu'on essaie de le faire par des interventions génétiques. Qu'on tente un jour, par exemple, d'implanter un gène P53 ou un équivalent à des gens qui sont menacés de cancers précoces, dans l'espoir d'augmenter leur espérance de vie. La difficulté, c'est qu'il s'agit de technologies de pays développés, qui risquent de renforcer d'autres inégalités. Que faire, dans ce cas: développer ces techniques au nom d'un principe d'égalité, ou y renoncer, pour éviter de créer des inégalités?

FJ – C'est une question importante. Il me semble que les deux arguments sont très différents, en réalité. Et il y a beaucoup d'enjeux sociaux dans ces questions.

AFL – Est-ce que par égalitarisme, nous devons choisir de renoncer à guérir nos tares parce que nous ne pouvons pas offrir les mêmes soins à tous? Cela rejoint les questions qui ont été agitées dans les périodes où l'on défendait l'eugénisme, où l'on rêvait de races supérieures, qui produiraient de beaux enfants etc., et aussi, en négatif, des races inférieures. En effet, imaginons aujourd'hui que dans les pays développés on puisse conférer aux gens un gène de protection au cancer. Il est clair qu'on ne pourra pas le faire pour les populations de l'Inde ou de l'Afrique, par exemple. Il y aura donc deux types de population.

FJ – C'est un raisonnement qui vaut dans beaucoup de cas.

AFL – Alors, faut-il se dire, comme dans le cas du SIDA, que dans un premier temps, les nouveaux traitements ont creusé les inégalités, mais que ce n'est qu'une période transitoire et que dans un second temps, tout le monde y aura accès. Comment voyez-vous les choses?

FJ – Au fond, la vraie question serait de savoir ce qu'en penseraient les gens, si on les consultait sur ce sujet.

AFL – Pour finir, que répondez-vous aux critiques qui s'en prennent à la généticisation de la médecine et qui reprochent à l'attitude analytique de ne plus considérer l'individu dans sa globalité, mais de se contenter de rechercher la pièce défectueuse, de faire la chasse aux mauvais gènes.

FJ – Je répondrai qu'il ne faut pas tout miser sur cette stratégie, mais qu'il faut s'en servir. Je ne crois pas qu'on puisse se priver de cette approche – et je ne vois pas l'intérêt qu'il y aurait à s'en priver. Mais ça ne doit pas être la seule approche. Il faut aussi appréhender les choses dans leur ensemble: la génétique n'est pas la voie unique. On ne peut pas se priver de la généticisation, mais ce n'est qu'un outil de travail.

AFL – Et sur la question de la procréation médicalement assistée et l'utilisation de techniques de génie génétique directement sur l'embryon?

FJ – J'ai toujours eu des réserves très fortes sur la PMA et sur la mise en avant du droit à l'enfant. Qu'il y ait un droit à la santé, pour les gens qui sont malades, c'est légitime. On doit faire ce que l'on peut pour améliorer leur sort. Sur le droit à l'enfant ou le droit à la PMA, je suis plus réservé. Et à partir du moment où on peut faire de la PMA, on risque de vouloir appliquer d'autres techniques sur les enfants. Où cela s'arrête-t-il? En outre, dans ce domaine particulier, on applique souvent les techniques sans protocole de recherche constitué, sans recherche préalable sur l'animal.

AFL – Oui, sans protocole, sans cahier d'expérience. Pensez-vous que c'est dangereux? Qu'il faudrait que ce soit interdit ou contrôlé?

FJ – Je crois qu'il faudra mettre un peu d'ordre, oui.

AFL – On pourrait choisir simplement de ne pas rembourser intégralement ces techniques?

FJ – C'est un procédé plus brutal, mais probablement plus efficace.

AFL – De toutes façon, il reste la question du droit à l'enfant. Vous pensez que la doctrine n'est pas bonne?

FL – Je pense qu'il faut faire tout ce qui est possible pour que les femmes puissent avoir des enfants quand elles le veulent, mais on ne peut pas le faire à n'importe quel prix.

AFL – Faut-il y voir un fond de respect de la nature?

FJ – Peut-être. J'ai été très frappé par des travaux qui montraient que la plupart des lésions du développement embryonnaire conduisaient à des avortements spontanés: c'est un formidable système naturel pour sélectionner ce qui marche. Bien sûr, on sait qu'il y a aussi des foetus normaux qui avortent, mais dans la grande majorité des cas d'avortements spontanés, il s'agit de fœtus anormaux. C'est un processus assez énigmatique. Qu'est-ce qui indique à l'ensemble du système que quelque chose ne va pas?

AFL – D'accord, c'est un argument pour respecter l'avortement spontané et ne pas forcer la poursuite de la grossesse quand il y a une menace d'avortement. Mais ce n'est pas une raison pour ne pas essayer de faire des enfants in-vitro.

FJ – Il faut essayer de donner satisfaction aux gens, mais il y a tout de même probablement des limites qu'on ne peut pas dépasser. On peut penser, par exemple, à la prématurité.

AFL – On peut s'interroger aussi sur l'innocuité de la procréation in-vitro: on s'est demandé si elle n'induisait pas des malformations ou des anomalies de développement. Les enquêtes qui ont été menées jusqu'à présent n'ont pas donné de résultat décisif. Mais dans les débuts de la procréation médicalement assistée, on se gardait bien de faire ces enquêtes.

En tout état de cause, vous pensez qu'il y a des limites naturelles – la prématurité, l'avortement spontané, etc. – et que la nature se régule d'elle-même. Il faut donc intervenir le moins possible. Pourtant, la médecine est très interventionniste. Et elle corrige la nature.

FJ – Il est vrai qu'elle a obtenu des succès.

AFL – Alors comment faut-il se comporter vis-à-vis de ce qu'on appelle la nature: faut-il la respecter ou doit-on constater qu'en réalité, la nature fait des erreurs qu'il nous appartient de corriger?

FJ – Oui, je pense que la nature fait des erreurs. Corriger ces erreurs, c'est sans doute un des fondements de la médecine. Quant à savoir s'il y a une règle à suivre pour savoir quand on doit intervenir ou non, c'est très difficile. On est plutôt dans des cas particuliers, à décider au coup par coup, comme dans les cas d'euthanasie.

AFL – Oui, là aussi, pour les fins de vie, la nature fait des erreurs. Acceptez-vous l'idée d'une «sagesse de la nature»?

FJ – Dans une certaine mesure.

AFL – Dans une certaine mesure, parce qu'en fait la nature improvise beaucoup, elle n'a pas de sagesse intrinsèque: simplement, les organismes trop déficients dégénèrent.

PART I

GENETICS AND THE LIFE SCIENCES

CHAPTER 1

GENETICS AND THE HUMAN LINEAGE

Can genetics throw some light on the evolution of the human lineage?

FRANCISCO JOSÉ AYALA[1], MIGUEL ÁNGEL CAPÓ[2],
CAMILO JOSÉ CELA-CONDE[2] AND MARCOS NADAL[2]

[1]*Department of Ecology and Evolutionary Biology, University of California, Irvine, USA fjayala@uci.edu*

[2]*Laboratory of Human Systematics, University of the Balearic Islands, Palma de Mallorca Spain, camilo.cela@uib.es*

Abstract: Genetics has greatly contributed to current knowledge about the evolution of the human lineage. Immunological techniques and the molecular clock have addressed the time divergence of human and chimpanzee lineages, disproving the late divergence model. The much-debated question about the origin of anatomically modern humans has been greatly by investigations of mitochondrial DNA, Y-chromosome DNA, and other genetic polymorphisms, all of which have also shed light on the size of human populations through hominid history. The persistent controversy concerning the relationship between Neanderthal and modern human populations has recently been constrained by the analysis of mitochondrial DNA fossil Neanderthals and early modern humans. Finally, recent genetic discoveries have contributed to unravelling the phylogenetic history of distinctive human traits, such as language and the masticatory apparatus

The main objective of this paper is to provide an overview of genetic contributions to the study of hominid evolution. We first review some traditional debates: (i) the evolutionary relationship between the great ape and human lineages; (ii) the debate between the "out of Africa" and "multi-regional" origin of modern humans; and (iii) the size of ancestral human populations. Thereafter we turn to recent genetic contributions concerning the relationships between Neanderthal and modern human populations, the evolution of language, and the reduction of the masticatory apparatus. We don't intend an exhaustive survey, but rather a brief look at a collaboration across disciplines that has been extremely fruitful, exploring traditional issues with a novel perspective and raising stimulating questions.

A. Fagot-Largeault et al. (eds.), The Influence of Genetics on Contemporary Thinking, 3–23.

1. RELATEDNESS BETWEEN GREAT APE AND HUMAN LINEAGES. MOLECULAR EVIDENCE

Based on biochemical analyses of blood serum proteins, Morris Goodman established in the early 1960s (Goodman et al., 1960; Goodman, 1962, 1963) that humans, chimpanzees, and gorillas are more closely related to each other than they are to orangutans. Goodman's measures of the relative immunologic distance among hominoids had profound taxonomic consequences. However, his studies did not estimate the time of divergence of various hominoid lineages. Sarich and Wilson (1967a, 1967b) used the "molecular clock" to conclude that the evolutionary branches leading to humans and chimpanzees diverged between four and five million years ago.

These early studies by Goodman and Sarich and Wilson are based on measures of immunological distance, which reflect differences between the amino acid sequences of proteins, but the correspondence between immunological distance and number of amino acid differences is not exact. Moreover, greater additional information can be obtained by examining DNA sequences, since not all nucleotide differences in the DNA result in amino acid differences between the encoded proteins. Time estimates based on direct comparison of DNA or protein sequences, as well as those based on immunological distance assume that there is a molecular clock, that is, that differences increase at a stochastically constant rate, which requires estimating the rate of change. This calibration of the molecular clock is achieved by reference to the fossil record. If an evolutionary event of divergence can be identified in the fossil record that occurred at an accurately determined time, the clock can be calibrated and used to time other divergence events. For example, if we know the time of divergence between Old World and New World monkeys, the number of molecular differences between the two lineages will allow to calibrate the molecular clock for a particular DNA segment or protein and determine the time of divergence between, say, humans and chimps based on their differences with respect to the same DNA segment or protein. As pointed out, among others, by Philip Tobias, this strategy implies a circular argument: we infer the timing of events evidenced in the fossil record based on rate parameters obtained from the fossil record itself (Tobias, 1991). Deviations from rate constancy and the vagaries of different calibrations have yielded estimates for the human and chimpanzee divergence ranging from 2.3 to 9.2 million years ago (Tobias, 1986).

In addition to proteins and DNA, chromosome number and structure may also be compared across lineages. Chaline (1991, 1996) combined their chromosome structure investigations with the available evidence obtained by molecular and immunological methods, concluding that the gorilla, chimpanzee, and human lineages separated almost simultaneously. The difficulty of resolving the order of divergence between the three lineages had been pointed out earlier by Goodman (1975), Bruce and Ayala (1979), and Smouse and Li (1987), among others, who suggested that a trichotomy might be assumed until further evidence would resolve the issue. Ruvolo's (1997) later review of the available DNA sequences strongly favored the divergence, first, of the gorilla lineage, soon followed by the divergence between the human and chimpanzee lineages. The close relationships among the ape lineages and the consistency required by the rules of cladistic taxonomy have led Goodman et al. (1998), to propose a radical classification (not universally accepted) such that the traditional family *Hominidae* divides into *Hylobatini* (common and siamang gibbons) and *Hominini*, and the latter into *Pongina* including *Pongo* and *Hominina,* which in turn includes the genera *Gorilla* and *Homo*. The genus *Homo* would then split into the subgenera H. (*Homo*) for humans and H. (*Pan*) for chimpanzees and bonobos.

2. THE ORIGIN OF MODERN HUMANKIND: "OUT OF AFRICA" OR "MULTIREGIONAL"?

The origin of anatomically modern humans is one issue that has sparked much paleoanthropological discussion. There is a consensus that modern humans evolved from hominids belonging to the *erectus* grade. The available evidence, however, is variously interpreted by different researchers specialized in Upper-Pleistocene hominids. There are two main hypotheses. The "multiregional" hypothesis suggests that the transition from *Homo erectus* to modern *Homo sapiens* took place concurrently in several regions of the Old World, involving several intermediate populations, and with frequent genetic exchange that kept the species' unity. The "Out of Africa" hypothesis states, on the contrary, that modern humans appeared somewhat before a hundred thousand years ago in Africa, from where they dispersed to the rest of the world. Previous populations of the rest of the world either disappeared without leaving descendants, or were replaced by modern humans migrating out of Africa. As pointed out by Clark (1997a, 1997b), the multiregional and out-of-Africa scenarios constitute paradigmatic positions, which allow for considerable latitude when evaluating available evidence.

Initial molecular data were interpreted as solid evidence favoring the "Out of Africa" hypothesis, albeit not in the extreme view held by some authors, according to which all modern humans descend from a single woman, the so-called "mitochondrial Eve". As we will expound later, this particular conception derives from confusion between the genealogy of genes and the genealogy of individuals. The "mitochondrial Eve" refers to the fact that the mtDNAs of all living humans have evolved from a single mtDNA, which coalesces between 100,000 and 200,000 years ago. But, it has been demonstrated that during millions of years the populations from which we descend were composed, on average, of at least 100,000 individuals (Ayala, 1995; Ayala and Escalante, 1996). If there ever was a population bottleneck, it never involved fewer than several thousand individuals, an issue to which we will return.

Advocates of the multiregional model underline what they interpret as fossil continuity in the transition from *Homo erectus* to "archaic" *Homo sapiens*, and thereafter to modern humans, in Australasia, the Middle East and other regions. These authors postulate that there were periodic genetic exchanges between populations of different regions, such that, despite geographical dispersion, the species evolved as a single genetic pool. Nevertheless, some geographical differentiation gradually emerged, which is currently reflected in genetic and morphological differences between ethnic groups (Wolpoff, 1989; Clark and Lindly, 1989; Thorne and Wolpoff, 1992; Bräuer, 1992; Clark, 1992; Waddle, 1994; Templeton, 2002). A problem with these claims is that there is no direct evidence of any related migrations as would be required for sustained genetic exchange between populations. Such negative evidence is not considered decisive by the supporters of the "multiregional" hypothesis.

The "Out of Africa" hypothesis proposes that anatomically modern humans evolved in Africa somewhat earlier than 100,000 years ago, dispersing there from the rest of the world, replacing any pre-existing human populations, whether *Homo erectus* or "archaic" *Homo sapiens*, in all regions (Cann, Stoneking, and Wilson, 1987; Stringer and Andrews, 1988; Stoneking et al., 1990; Vigilant et al., 1991; Stringer, 1992; Ruvolo et al., 1993; Cavalli-Sforza, Menozzi and Piazza, 1994; Goldstein et al., 1995; Horai et al., 1995; Rogers and Jorde, 1995). Traces of hybridization between modern humans arriving from Africa and pre-existing human populations have never been found in Eurasia, although this does not necessarily imply that hybridization never occurred (see below, Chapter 9).

The reconstruction of the mtDNA genealogical tree places its roots, that is, the origin of ancestral mtDNA, in Africa (Cann et al., 1987; Stoneking et al., 1990; Vigilant et al., 1991; Ruvolo et al., 1993; Horai et al., 1995). Early mtDNA studies focussed on the control region, which represents less than 7% of all mitochondrial genetic information and does not have a coding role. A study of the complete mtDNA (16,500 base pairs) from 53 individuals has confirmed the same African origin (Ingman et al., 2000). The mtDNA evidence would not be conclusive by itself, given that mitochondrial DNA constitutes a tiny fraction of the total human DNA, which at three billion base pairs is 400,000 times larger than the mtDNA. But chromosome DNA microsatellites (Goldstein et al., 1995) and of a large sample of nuclear genes spread throughout the entire human genome (Cavalli-Sforza et al., 1994) also yield genealogical trees rooted in Africa.

Ancestral African populations appear on these genealogical trees set apart from all non-African populations, which are located on a single branch emerging from the multibranched African tree. The most profound divergence of non-African populations in this genealogical tree is calculated at 156,000 years ago (with a possible error of tens of thousands of years), which would mark the earliest possible point in time at which modern humans would have dispersed from Africa to the rest of the world. Ethnic differentiation among modern populations would be a relatively recent event, a result of diverging evolution among populations separated only for the last 50,000 or 100,000 years. This conclusion, emerging from the genealogical trees, is consistent with extensive studies of genetic polymorphism, showing that living human populations from different parts of the world are not greatly differentiated.

Advocates of the multiregional hypothesis have recently presented supporting molecular data. The mtDNA sequenced from ten fossil specimens of anatomically modern humans, retrieved from two different regions of Australia, most dated 2,000–15,000 years old, but one, LM3, around 60,000 years old, have shown a mtDNA sequence in specimen LM3 which is absent from the other ancient specimens, as well as from present-day modern humans (Zimmer, 1999). The inference is that the genetic diversity of this sample is much higher than expected under the scenario of a recent modern-humans origin, thereby supporting the multiregional hypothesis. However, alternative interpretations have been suggested. Svante Pääbo (cited by Holden, 2001), believes that the Australian team failed to maintain the necessary precautions for avoiding contamination, and moreover that what the results really show is that the polymorphism of anatomically modern humans is higher than previously thought. According to Pääbo, the African origin and replacement scenario remains the most plausible one.

Templeton (2002) has supported the multiregional hypothesis by proposing a scenario with not just one, but several migrations out of Africa. There would have been a mixture of populations rather than a single population substitution, yielding a predominance of immigrant genomes proceeding from Africa, given the advantageous modern human genotype evolved there. However, there is not any supporting palaeontological evidence, nor any other sort of evidence for migrations as early as those proposed by Templeton (400,000 years ago), nor have any recent studies uncovered the suggested morphological continuity in Asia or in other continents (Cavalli-Sforza, 2003).

Studies of human genetic polymorphism have revealed information consistent with a recent origin of all living human populations, as proposed by the out-of-Africa hypothesis. When the genetic diversity of human populations is mapped out geographically, it is found that 85% of it is present in any local population, this is to say, in any village or city of any continent (although the genes contributing to this 85% vary from one population to another). From 5 to 6% additional genetic variation is found when local populations on the same continent are compared, and an additional 10% when populations from different continents are compared (Barbujani et al., 1997; Jorde et al., 1997; Kaessmann et al., 1999). This seems at first surprising when we consider how easy it is to distinguish a Congolese, a Swede and a Japanese. The explanation is that ethnic differences, such as the color of the skin and other observable morphological features, are associated with a small number of genes, which became differentiated because of their high adaptive value in response to different latitudes and climates. The distribution of genetic variation among populations does not give the time of dispersion of modern humans throughout the world, but indicates that the dispersion could not be very ancient, given the relatively little genetic differentiation existing among continents.

3. THE MITOCHONDRIAL EVE AND *ZFY* ADAM

As we have explained, most of the genetic information is found in the chromosomes, whereas the amount of DNA in mitochondria is relatively small and follows a matrilineal inheritance pattern. We have also mentioned that modern human mtDNA sequences coalesce in an ancestral sequence, known as mitochondrial Eve, present in Africa about 200,000 years ago (Cann et al., 1987; Stoneking et al., 1990; Vigilant et al., 1991). This Eve, however, is not the only woman from which all present day humans descend, but a mtDNA molecule from which all current mtDNA molecules descend.

The inference that all women descend from a single or very few women (Brown, 1980; Lowenstein, 1986) stems from the confusion between gene genealogy and a genealogy of individuals. This may be illustrated with an analogy. A present-day surname can be shared by many people in different continents, but it may have a singular origin centuries ago. If we accept that the surname is transmitted only from father to sons, all those carrying the surname will be descendants, by paternal line, from the "founder", the family's "Adam", but those people will also descend from many other men and women who lived before and after the founder. Similarly, many contemporary women to the mitochondrial Eve have left descendants in modern humanity, contributing with nuclear genes.

The legitimate conclusion of the mtDNA analysis is that the mitochondrial Eve is the matrilineal ancestor of modern humans. Everyone has a single matrilineal ancestor in any given generation. Everyone inherits their mtDNA from their great-grandmother through maternal lineage, but they also inherit other genes from the other three great-grandmothers and from their great-grandfathers. The mtDNA we inherit from the mitochondrial Eve represents 1/400,000 of our total DNA. The rest of DNA has been received from other individuals contemporary or not of the mitochondrial Eve.

The coalescence of the mtDNA of modern humans into a single ancestor is a feature that necessarily occurs for any one gene or genetic trait. As one proceeds back in time, at any gene locus (or DNA segment) all 2N genes of a species with N individuals derive from fewer and fewer ancestral genes, eventually converging into a single gene ancestor to all 2N descendants. But the ancestral genes for different gene loci occur in different generations and, of course, different individuals. The genome of each living human individual derives from many ancestors. The converse of this is the non-intuitive inference that any human who lived a few thousand generations ago and who has living descendants is an ancestor of all living individuals (Rohde et al., 2004; Hein, 2004), although he/she would have contributed different genes to different living individuals.

The Y chromosome is the genetic counterpart of mtDNA in that it is inherited only from fathers to sons. There are regions on chromosome Y that are not homologous to chromosome X and thus are transmitted only through the paternal line. A DNA fragment of 729 nucleotides of the *ZFY* gene (probably involved in testicle or sperm maturation) found on chromosome Y was sequenced, in 38 men representative of the main ethnic groups, by Dorit and colleagues (1995). The theory of coalescence leads these authors to the conclusion that the origin of modern human Y chromosomes dates back to a Y chromosome close to 270,000 years ago, with a confidence margin extending from zero to 800,000 years. The ancestral Adam from whom all living men have inherited the Y chromosome was not, however, our only male ancestor in his own or any other generation. As in the case of the mitochondrial DNA, the rest of our genes come from many other different male and female ancestors (Ayala, 1995).

Cavalli-Sforza and colleages (1994) have pointed out that discrepancies between the calculated bifurcation time between African and non-African populations based on nuclear genes (about 100,000 years ago) and mtDNA (close to 200,000 years ago) are to be expected. Divergence time estimates in such studies show great variation, largely due to the limited data set they are based on. It is unsurprising, therefore, that mtDNA polymorphism coalescence has been estimated at 143,000 years by Horai and colleagues (1995) and at 298,000 years by Ruvolo and colleagues (1993), with confidence intervals ranging from 129,000 to 536,000 years. The differences between estimates based on mtDNA, Y chromosomes, and other nuclear genes are also due to gender and social differences in migration patterns (Cavalli-Sforza and Feldman, 2003). For example, patrilocal marriage has historically been more common than matrilocal, which can explain differences between mtDNA and Y-chromosome patterns in different populations. Demographic differences between the sexes, such as greater male than female mortality, the greater variance in reproductive success of males than females, and possibly the greater frequency of polygyny than polyandry, may explain the discrepancy between estimated dates obtained form the non-recombining part of the Y chromosome and from mtDNA.

4. SIZE OF ANCESTRAL HUMAN POPULATIONS

Molecular evolution data favor an African origin for modern humans, but there is no reason to assume that a severe population bottleneck occurred at the time of origin of modern humans. As we have pointed out above, the "mitochondrial Eve" notion that modern humans descend from a single woman who lived in Africa between 100,000 and 200,000 years ago, is a mistake associated with the confusion between the genealogy of genes (which leads to fewer and fewer ancestral genes as one goes farther and farther back in the genealogy) and the genealogy of individuals (which increases by a factor of two in each previous generation – two parents, four grandparents, etc. – although eventually the same ancestors appear time and again in the genealogy). The ancestral number of individuals in each generation can be estimated studying very polymorphic current genes by means of the genetic theory of coalescence, such as the immune system's main histocompatibility complex (MHC).

The amino acid composition varies from one MHC molecule to another, which is accounted for by the extensive polymorphism that characterizes MHC genes. In human populations, as in other mammals, there are a great number of genetic variants (alleles) in any of the different MHC loci, and the alleles may differ amongst themselves in up to 100 nucleotides (Marsh and Bodmer, 1991; O'Huigin et al., 1993; Bontrop, 1994; McDevitt, 1995). The MHC polymorphisms are very ancient, with genetic families that can be traced millions of years back through the primate lineage (Gyllenstein and Erlich, 1989; Gyllenstein, Lashkary and

Erlich, 1990; Ayala et al., 1994; Bergström and Gyllenstein, 1995; Ayala and Escalante, 1996).

The method used to reconstruct the evolution of DNA may be illustrated with a simple example. If we compare two particular human *DRB1* gene sequences (*Hs*1103* and *Hs*0302*) with two chimpanzee *DRB1* gene sequences (*Pt*0309* and *Pt*0302*), each composed of 279 nucleotides, it is the case that the human gene *Hs*1103* is more similar to the chimpanzee gene *Pt*309*, than to human *Hs*0302*, which in turn is most similar to the chimpanzee *Pt*0302* (see Table 1). It follows that the lineages of both human genes diverged from each other before each diverged from one of the chimpanzee genes; and, therefore, more than six million years ago, the approximate time humans diverged from chimpanzees (Figure 1). The early origin of these and other *DRB1* lineages is the feature that makes them particularly adequate for the study of early human populations.

Figure 2 (see below) represents the genealogy of 119 *DRB1* genes. Fifty-nine of them are human, 40 of them belong to apes, and 20 of them to Old World monkeys. The length of each branch is proportional to the number of nucleotide substitutions that have occurred along it, and the degree of relatedness among genes is reflected in the closeness of branches. For example, nine human genes in the upper part of Figure 2 are closely related among themselves, but more remotely related with other human genes than with the six genes immediately below them in the figure, which include genes from other primate species. The divergence of the lineage

Table 1. Nucleotide differences between four human and chimpanzee *DRB1* genes

Gene	*HS*1103*	*Pt*0309*	*Pt*0302*
*HS*0302*	18	18	12
*Hs*1103*		9	20
*Pt*0309*			20

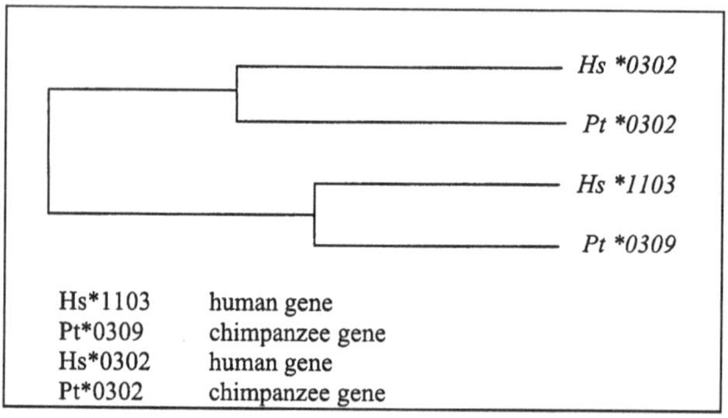

Figure 1. Genealogic tree of four *DRB1* genes, with their evolutionary divergence relationships. The length of the branches is proportional to the difference in nucleotides between the genes

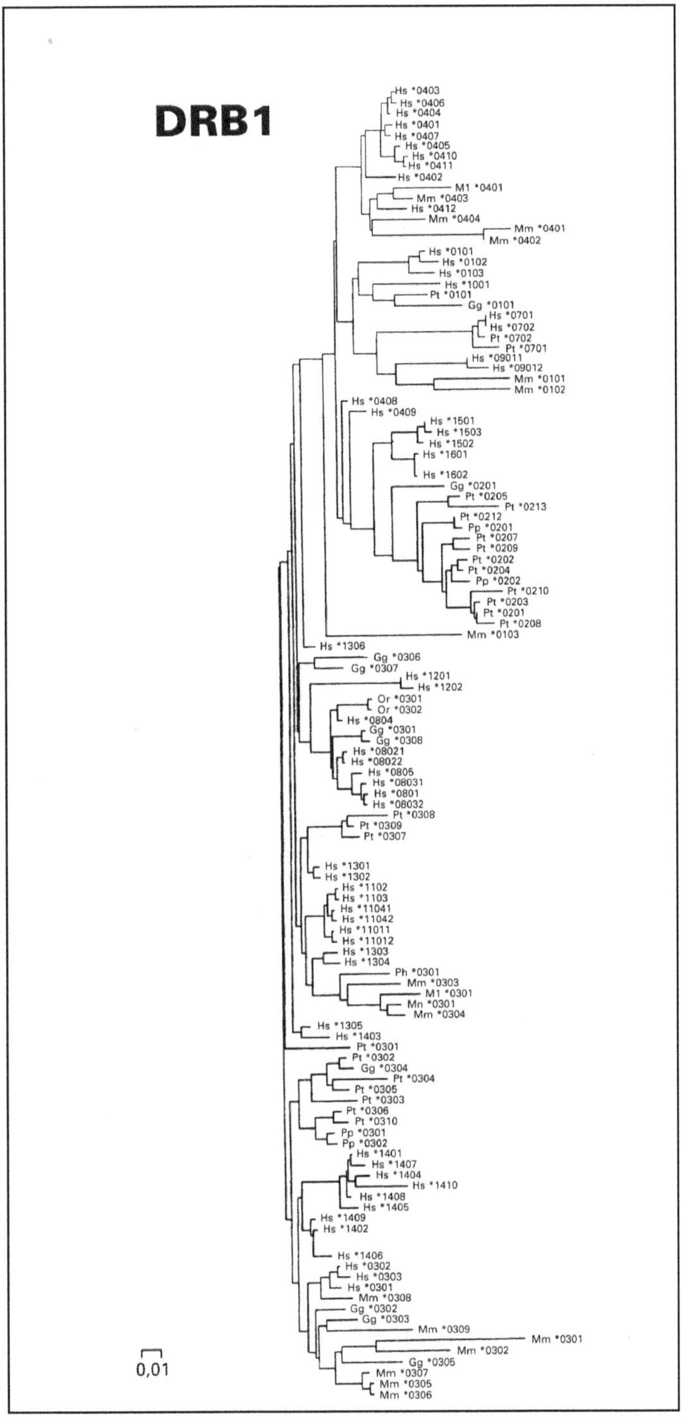

Figure 2. Genealogic tree of 119 human and other primate *DRB1* (exon 2) genes

leading to New World Monkeys and the lineage leading to Old World Monkeys and Hominoids occurred close to 35 million years ago, at the border between the Eocene and the Oligocene, so the relationships represented in Figure 2 manifest that several human gene lineages already existed at that time. The age of the human *DRB1* lineages can be calculated, additionally, by calibrating the *DRB1* molecular clock, which changes at a substitution rate of 1.06×10^{-9} per site per year (rate obtained by the method of the "minimum" described in Ayala and Escalante, 1996).

Figure 3 (see below) represents the genealogy of 59 human *DRB1* genes, using the mean rate of evolution to determine branch length. The timescale at the base represents three particular events in human evolution: the divergence of the orangutan lineage 15 million years ago, the divergence of humans and African apes (chimpanzees and gorillas) some six million years ago, and the appearance of *Homo erectus* represented at 1.7 million years ago. These reference points are useful to determine the number of genetic lineages existing at any given point in time, which is done by counting the number of lineages intersected by a vertical line drawn from the desired point in time represented on the X-axis. For instance, close to 6 million years ago, 32 current human genetic lineages already existed. The genealogy of all human genes coalesces near 60 million years ago, which means that the set of human *DRB1* genes present in our species started diverging that long ago. If 32 lineages of the *DRB1* gene have persisted since 6 million years ago, it is evident that at any given moment thereafter, there were no less than 16 individuals. The minimum number must in fact be much larger, because the probability of those 16 individuals each carrying two different alleles and these being, in turn, different to the rest of the alleles carried by other individuals is zero. We can achieve more precise estimates of population size by means of gene coalescence theory, as well as by other recently formulated mathematical theories. Ayala (1995), Ayala et al. (1994), and Ayala and Escalante (1996) have applied gene coalescence theory to the *DRB1* and *DQB1* genes, concluding that the age of the polymorphisms indicates that the mean effective size of early human populations (i.e., the number of individuals that have direct descendants among present-day humans) has not been smaller than about 100,000 individuals.

This estimate must be further assessed by taking into account two features of coalescence theory. Firstly, the average size of populations is calculated using the *harmonic* mean, a value which is consistent with much higher values than 100,000 in one or several generations, but it is not consistent with values that are much smaller. Secondly, the mean refers to the *effective* number of individuals, those who are capable of reproducing at any given time. The *census* number, that is, the total population, is roughly four or five times larger than the effective number. Thus, on average, the census number of early human populations would have been between 400,000 and 500,000 individuals.

The conclusions reached with coalescence theory have been confirmed by means of experiments carried out on computers. In these experiments "populations" are collated that consist of a given number of individuals, which reproduce according to human population norms and are maintained for as many thousands of years

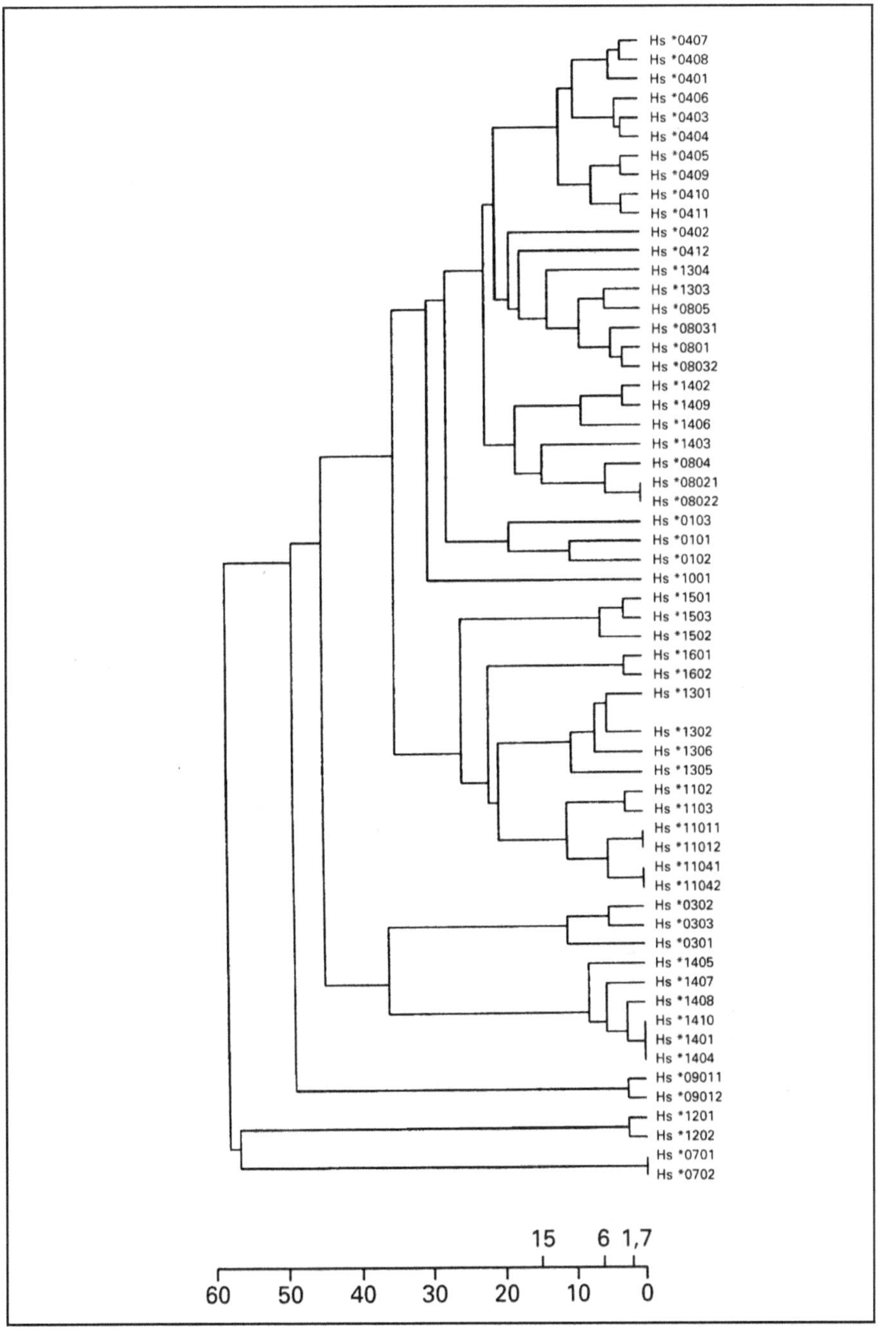

Figure 3. Genealogical tree of 59 human DRB1 genes, constructed according to the average linkage method, which assumes equal rates across all branches (The branch length scale is millions of years)

as desired. It has been confirmed that in order for actual humanity to possess 59 alleles of the gene *DRB1*, human ancestral populations must have consisted, on average, of at least 100,000 individuals during the last 2 million years, the same number obtained by means of coalescence theory (see particularly Ayala, 1995; Ayala and Escalante, 1996). The minimum number of reproductive individuals who could have lived at any one time would have never been smaller than between 4,500 and 10,000 individuals (Ayala, 2000).

5. RECENT CONTRIBUTIONS AND DEVELOPMENTS

5.1. Neanderthals and Modern Humans

As mentioned above, the debate concerning modern human origins has been influenced by implicit assumptions regarding what it means to be human. The recent retrieval of mtDNA from Neanderthal and Cro-Magnon specimens (Krings et al., 1997, 2000; Ovchinnikov et al., 2000; Schmitz et al., 2002; Serre et al., 2004) constitutes a landmark in the history of the debate on our species' origin, because mtDNA sequences can be used to test the multiregional and out-of-Africa hypotheses. Investigation of fossil mtDNA is hindered by the scarcity of valid samples, ignorance of ancient population dynamics, the difficult technical procedures involved, and the possibility of contamination with modern human DNA. Authors supporting multiregional models (Relethford, 2001; Templeton, 2002; Wolpoff et al., 2000) have claimed that Neanderthals and anatomically modern humans are two populations of a single species. As stated by Wolpoff et al. (2000), modern humans emerged from the occasional merging and splitting of multiple populations over millions of years, with extensive genetic exchange. This scenario could be supported by finding mtDNA sequences shared by (early) modern humans and Neanderthals.

In spite of some claims to the contrary, there is no valid evidence of any mtDNA gene flow between Neanderthal and early modern humans (Serre et al., 2004). However, the conclusion is not warranted that gene flow never occurred. As Nordborg (1998) has pointed out, the possibility exists that modern human populations may have lost any Neanderthal mtDNA through random drift. Assuming that human populations have consisted of 10,000 reproductive individuals per generation, it is possible, nevertheless, to exclude that the amount of DNA coming from Neanderthal ancestors would have been greater than 25% of the total.

It is possible to rule out completely that Neanderthals contributed to the genetic making of our species? Not presently. Sequencing additional ancient DNA from Neanderthal and early modern human specimens could confirm the presence of Neanderthal DNA in modern humans, but presently one cannot exclude the possibility that 25% or less DNA would have come from the Neanderthals, but most of it was lost through the generations by random drift. In any case, "while it cannot be excluded that Neanderthals contributed variants at some genetic loci to contemporary humans, no positive evidence of any such contribution has yet been detected" (Serre et al., 2004).

5.2. Language

Language is one of our species' most notable traits. Multidisciplinary research on language in the last forty years has changed our understanding of human cognition. The evolution of human language remains, however, largely unresolved. The impressive growth of research on language evolution made some headway, but has not yet produced satisfactory understanding of how the capacity for language came about. Recent work has, on the one hand, challenged traditional assumptions and pointed out additional variables that need to be taken into account. For instance, Daniel Lieberman and colleagues (Lieberman and McCarthy, 1999; Lieberman et al., 2001) have challenged the assumption (Lieberman et al., 1972) that phonation constraints are virtually the sole determinants of the modern human vocal tract configuration. They have shown how factors relating to swallowing, respiration, and facial ontogenetic growth must also be taken into account in order to obtain an accurate picture of the evolution of speech and language. On the other hand, recent research has also helped to characterize the entity that evolved, that is, language, with greater accuracy. For instance, Fitch and Hauser (2004) have revealed that other primates can master some simple grammars, but not certain types of grammatical constructions, such as phrase structure grammar. Results of this sort suggest that human language is a combination of novel and primitive language-related capacities. Separating the true novelties from primitive characters should enable researchers to define which specific traits of language appeared during human evolution.

Noam Chomsky, despite his scepticism regarding the efforts to answer the question of language evolution, revitalized the idea of language as an innate faculty, challenging the mainstream behaviorist tradition (Chomsky, 1966). Since then many researchers have assumed that there are certain human genes related only to language, although only recently, has the function of a specific gene been related to language (Lai et al., 2001). This gene, named *FOXP2,* was isolated while studying a family with an inherited severe language and speech impairment. The affected members of the family have difficulties in selecting and sequencing fine orofacial movements, and manifest grammatical deficits as well as slight nonverbal cognitive impairments. It was soon determined that the disorder is transmitted as an autosomal dominant monogenetic trait, associated with a point mutation in a gene situated on chromosome 7 (Lai et al., 2000). Brain imaging techniques have revealed functional and structural differences between affected members of the family and unaffected members and control subjects (Liégeois et al., 2003; Vargha-Khadem et al., 1998; Watkins et al., 2002). These differences vary slightly according to the particular imaging technique used and the task performed during the procedure, but the most consistent result is an association with a subcortical region, the caudate nucleus, the volume of which is reduced bilaterally in the affected members of the family (Watkins et al., 2002). This observation is striking, because the traditional viewpoint tends to restrict language-related areas to the cortex. It seems that *FOXP2* is involved in the regulation of the development of subcortical neural circuitry critical for language and speech.

FOXP2 is not exclusively human, but is part of the genome of animals as evolutionarily distant as humans and mice. Surprisingly, the *FOXP2* protein is highly conserved among mammals: it has undergone only one amino acid replacement during 70 million years or so that elapsed between the last common ancestor of primates and mice and the last common ancestor of humans and chimpanzees. Since the divergence of the human and chimpanzee lineages, nearly 8 million years ago, the human protein has undergone two amino acid changes, while the chimpanzee form has not changed. If the gene participates in laying down the neural circuits involved in speech and language, it seems possible that the last two mutations that occurred in the human lineage were crucial for the development of language. Enard and colleagues (2002) have suggested that these events happened during the last 200,000 years. We do not know whether language would have been possible with a chimpanzee or earlier hominid version of the gene, but the estimated age of the mutated gene fits well with the estimated age for the appearance of modern humans. One possibility is that language appeared more-or-less suddenly in the hominid lineage with the advent of modern *Homo sapiens*. However, the high degree of conservation of the protein, and of the pattern of the gene's expression in the brain, suggests that language and speech are, at least in part, supported by neural structures present in other species, which would support a gradual emergence of the capacity for language through the recruitment or fine tuning of pre-existing neural pathways.

There are some problems with this genetic approach to language evolution. The specific deficits affecting the studied family are not clear, and their importance varies according to different authors. The assessments range from grammatical problems (Gopnik and Goad, 1997) to motor sequencing defects (Vargha-Khadem et al., 1995). There is no unanimity regarding which aspect of language is impaired by the mutation; whether several aspects are affected; or whether non-linguistic traits are also affected. In any case, the mutation does not produce a complete impairment of language. At present, it is not known what is the function of the normal gene. However, as Bishop (2002) has pointed out, given that that the gene product is a transcription factor, *FOXP2* probably plays a key role in the regulation of other genes.

A relevant consideration is that *FOXP2* is expressed in tissues other than the brain during embryonic development, including the lungs, gut and heart, as well as in several adult tissues (Shu et al., 2001). This is consistent with what is known about other transcription factors, many of which seem to impact diverse functions. Lai et al. (2003) have pointed out that the homologous expression of the gene in humans and mice indicates that this gene has an important role in the development of motor circuits in mammals. This implies that the evolution of brain features related to language may have involved more the developmental modification of pre-existing motor-related features, rather than the addition of new features. Thus, current knowledge about the genetics of language indicates that the existence of specific and exclusive language genes is unlikely. The neural mechanisms that give rise to language seem to be the result of the interaction of several genetic cascades.

Language is not a monolithic faculty, independent from other brain/mind features, but is rather the result of mechanisms operating in parallel and in close relation to other features.

The *FOXP2* findings suggest that (i) language did not appear suddenly, but is grounded in neuronal circuitry involved in cognition and motor control, and its evolution should be studied accordingly; (ii) the evolution of language does not depend on the creation of new brain areas, but rather it is most likely associated with the fine wiring of pre-existing brain structures; (iii) the relation between genes and language is bound to be more complex than previously suspected, with many genes interacting with one another, and participating in the regulation of other functions in various parts of the body, so that single mutation or abrupt change scenarios of language evolution (Berwick, 1998; Bickerton, 1995) would seem implausible.

5.3. Masticatory Apparatus

The early hominids retained ape-like traits, but even the earliest ones exhibited derived traits. The tendency towards gracile hominids began close to 3.5 million years ago, but a noteworthy change occurred in the hominid lineage about a million years later. McHenry and Coffing (2000) have pointed out some features involved in this transition:

> Before about 2.6 mya, stone tools were absent at sites containing hominid fossils, brain sizes were chimp-like, cheek teeth and supporting masticatory structures were enormous, numerous primitive traits were retained in all parts of the body, including the skull, bodies were small, there was strong sexual dimorphism in body size, and hindlimbs were small relative to forelimbs. By 1.8 mya, *H. ergaster* stepped into view with its more human-like body and behavior.
>
> (McHenry and Coffing, 2000)

Unfortunately the hominid fossil record is not definitive about the specific configuration and timing of this landmark conglomerate of events. Specifically regarding the masticator apparatus, which is one of the most informative traits for identifying fossil remains, the picture is far from clear: "The size of the cheek teeth and other parts of the chewing apparatus reduced in *H. habilis* perhaps as early as 2.3 mya, but certainly by 1.8 mya" (McHenry and Coffing, 2000). A fair amount is known about the australopithecine's diet, obtained from a number of fossil indicators. Their diet consisted mainly of hard vegetable fibers, such as tubercles, roots, and plant stems. It is also known that *Homo erectus* increased the amount of meat in its diet, as it is inferred from the reduction in robusticity of morphological elements associated with mastication. The transition from robust to gracile hominids is one of the most complex and interesting episodes in the evolution of our lineage. It is now clear that gracile forms did not simply replace robust ones, or that teeth and jaws just became smaller. Rather, a number of distinct gracile and robust hominids overlapped temporally and spatially. Moreover, some species, such as *Homo habilis*, reveal a mosaic of primitive and derived traits, and its classification as gracile or robust, or indeed as *Homo* or *Australopithecus*, is open to interpretation (Wood and Collard, 1999).

Given the incomplete and ambiguous picture of the initial stages of the genus *Homo* painted by fossil remains, the introduction of a genetic perspective concerning the chewing apparatus by Steadman and colleagues (2004) is very welcome. These authors have identified a novel exemplar of the myosin heavy-chain coding genes. Myosin proteins are a key element in the generation of muscle contractile force. (Currie, 2004). The inactivation of some of these genes results in an important decrease in the size of the muscles in which they are expressed. The specific gene identified by Steadman et al. (2004), known as *MYH16*, is mainly expressed in primates in the muscles that move the jaw. The human version of this gene presents a mutation which prevents the *MYH16* protein from concentrating in the cells. This genetic characteristic is reflected in a clear difference between humans and other primates in the robustness of the muscles involved in chewing. An additional issue, relevant for palaeoanthropology, is that the human inactivating mutation became fixed about 2.4 million years ago, just before the time when fossil specimens appear that are uncontroversially included in the genus *Homo*. When compared with earlier hominids, *Homo erectus* fossils are characterized by a relatively gracile masticatory apparatus. Steadman et al. (2004) hypothesize that the *MYH16* human mutation led to a reduction in the size and contractile force produced by masticatory muscles, and that this decrease translated into a reduction of bony structures related with mastication, such as the sagittal crest or the flaring of zygomatic arches. Additionally, this decrease in muscle size removed an evolutionary constraint on brain size (Steadman et al., 2004). Once the stress produced by masticatory muscles and their bony fixation points were reduced, the brain and the braincase were able to grow under the appropriate selective pressures.

The plausible picture of human evolution just described, needs to be set in a wider framework. Around the time the *MYH16* mutation occurred, other changes were taking place in some hominid lineages, such as the overall reduction in teeth size, enamel thinning, restructuring of the brain, and the construction of simple stone tools. It is clear that *Homo erectus* is very different from any *Australopithecus*, but their differences stem not only from the robusticity of masticatory muscles, but also include the appearance of modern limb proportions, increase in height and weight, increase in brain volume, reduction in size as well as the appearance of new adaptive strategies which involved the construction of sophisticated stone tools, hunting strategies, and the colonization of lands far from Africa. The reduction of musculature and bone structures associated with heavy mastication seems to be a piece of a mosaic of changes, although it is not yet clear how all the changes are interrelated.

6. CONCLUDING REMARKS

The traditional study of human evolution has been based on hard-tissue phenotypic traits susceptible to fossilization, in addition to tools and other artefacts. This approach is hampered by the fact that sometimes evidence is incomplete, contradictory, or ambiguous, and susceptible of different interpretations. Furthermore, information regarding soft tissue and functional traits must be indirectly inferred from the fossil and archaeological record. In spite of these difficulties, traditional paleoanthropological sources have yielded a considerable wealth of knowledge regarding the evolutionary events that shaped our lineage. Alternatively, molecular biology and genetics have begun exploring a novel level of evidence of human evolution. By means of plausible inferences based on the human genotype, they are able to offer information about the way it might have attained its present form. We believe that the identification of genetic correlates of evolutionary episodes identified on the basis of phenotypic remains studied by palaeoanthropologists, as well as the incorporation of genetic data as hypotheses constraints, represent fruitful challenges derived from the increasing influence of molecular biology and genetics on paleoanthropology.

In the present work we have attempted to illustrate how genetics has helped shape the answers to some of the most relevant questions in paleoanthropology through two kinds of examples. Firstly, population genetics has been fundamental in the detailing of two major events in human evolution: the hypothesis of the late divergence between human and chimpanzee evolutionary lineages, and the recent African origins of modern humans. Secondly, the study of the genetic correlates of specific morphological traits and hereditary deficits of certain cognitive capacities has allowed the formulation of hypothesis regarding their phylogenetic history. Despite that genetics need not have the last word in this kind of discussions, palaeoanthropological research is greatly enriched with this additional perspective.

ACKNOWLEDGEMENTS

Miquel Capó and Marcos Nadal acknowledge support by the Govern de les Illes Balears (FPI-03-41498911-Q9 and FPI-01-43098605-D, respectively).

REFERENCES

Ayala F (1995) The myth of eve: molecular biology and human origins. Science 271:1930–1936

Ayala F (2000) ¿Desde África? Una perspectiva de los elementos poblacionales en la aparición de los seres humanos modernos. Ludus Vitalis VIII (13):134–156

Ayala F, Escalante A (1996) The evolution of human populations: a molecular perspective. Mol Philos Genet Evol V(1):188–201

Ayala F et al. (1994) Molecular genetics of speciation and human origins. Proceedings of the National Academy of Sciences of the United States of America 91:6787–6794

Barbujani G et al. (1997) An apportionment of human DNA diversity. Proceedings of the National Academy of Sciences of the United States of America 94:4516–4519

Bergström T, Gyllenstein U (1995) Evolution of Mhc class II polimorphism: the rise and fall of class II gene function in primates. Immunol Rev 143:13–32

Berwick RC (1998) Language evolution and the Minimalist Program: the origins of syntax. In: Hurford JR, Studdert-Kennedy M, Knight C (eds) Approaches to the evolution of language. Cambridge University Press, Cambridge

Bickerton D (1995) Language and human behavior. University of Washington Press, Seattle

Bishop DVM (2002) Putting language genes in perspective. Trends Genet 18:57–59

Bontrop RE et al. (1995) Evolution of major histocompatibility complex polimorphism and T-cell receptor diversity in primates. Immunol Rev 143:33–62

Bontrop RE (1994) Non human primate Mhc-DQA and DQB second exon nucleotide sequences: a compilation. Immunogenetics 39:81–92

Bräuer G (1992) Africa's place in the evolution of homo sapiens. In: Bräuer G, Smith FH (eds) Continuity of replacement? controversies in *homo sapiens* evolution. Rotterdam, Balkema, pp 83–98

Brown WM (1980) Polimorfism in mitochondrial DNA of humans as revealled by restriction endonuclease analysis. Proceedings of the national academy of sciences of the United States of America 77:3605–3609

Bruce EJ, Ayala FJ (1979) Philogenetic relationships between man and the apes: electrophoretic evidence. Evolution 33:1040–1056

Cann RL (1988) DNA and human origins. Ann Rev Anthropol 17:127–143

Cann RL, Stoneking M, Wilson AC (1987) Mitochondrial DNA and human evolution. Nature 325:31–36

Cavalli-Sforza L, Feldman M (2003) The application of molecular genetic approaches to the study of human evolution. Nature genetics supplement 33:266–275

Cavalli-Sforza LL, Menozzi P, Piazza A (1994) The History and Geography of Human Genes. Princeton University Press, Princeton, NJ

Chaline J et al. (1991) Un modèle chomosomique et paléobiogéographique d'évolution des primates supérieurs. Géobios 24:105–110

Chaline J et al. (1996) Chromosomes and the origin of apes and australopithecines. Hum Evol 11:43–60

Chomsky N (1966) Cartesian linguistics. Harper & Row, New York

Clark GA (1997a) Through a glass darkly: conceptual issues in modern human origins research. In: Clark GA, Willermet CM (eds) Conceptual issues in modern human origins research. Aldine de Gruyter, New York

Clark GA (1997b) The middle-upper paleolithic transition in Europe: an American perspective. Norw Arch Rev 30:25–53

Clark GA (1992) Continuity or replacement? putting moderns human origins in an evolutionary context. In: Dibble H, Mellars P (eds) The middle paleolithic: adaptation, behavior and variability. University Pennsylvania Museum, Philadelphia, pp 183–205

Clark GA, Lindly JM (1989) The case for continuity: Observation on the biocultural transition in Europe and Western Asia. In: Mellars P, Stringer C (eds) The human revolution: behavioral and biological perspectives on the origins of modern humans. Princeton University Press, Princeton, NJ pp 626–676

Currie P (2004) Human genetics: Muscling in on hominid evolution. Nature 428:373–374

Dorit RL, Akashi H, Gilbert W (1995) Absence of polimorphism at the ZFY locus on the human Y chromosome. Science 268:1183–1185

Enard W, Przeworski M, Fisher SE, Lai CSL, Wiebe V, Kitano T, Monaco A P, Pääbo S (2002) Molecular evolution of FOXP2, a gene involved in speech and language. Nature 418:869–872

Fitch WT, Hauser MD (2004) Computational constraints on syntactic processing in a onhuman primate. Science 303:377–380

Goldstein DB et al. (1995) Microsatellite loci genetic distances, and human evolution. Proceedings of the National Academy of Sciences of the United States of America 92:6723–6727

Goodman M (1962) Evolution of the immunologic species specificity of human serum proteins. Hum Biol 34:104–150

Goodman M (1963) Man's place in the philogeny of the primates as reflected in serum proteins. In: Washburn SL (ed) Classification and human evolution. Aldine, Chicago, IL, pp 204–234

Goodman M (1975) Protein sequence and immunological specifity. In: Luckett WP, Szalay FS (eds) Philogeny of the primates. Plenum Press, New York, NY, pp 321–353

Goodman M, Poulik E, Poulik MD (1960) Variations in the serum specifities of higher primates detected by two-dimensional starch-gel electrophoresis. Nature 188:78–79

Goodman M, Czelusniak J, Page S, Meireles C (1998) Where DNA Sequences place homo sapiens in a phylogenetic classification of primates. In: Tobias PV, Raath MA, Moggi-Cecchi, JY, Doyle GA (eds) Humanity from African naissance to coming milennia. Firenze University Press & Witwatersrand University Press, Firenze, Italia

Gopnik M, Goad H (1997) What underlies inflectional error patterns in genetic dysphasia? J Neurolinguistics 10:109–137

Gyllenstein UB, Erlich HA (1989) Ancient roots for polimorphism at the HLA-DQ locus in primates. Proceedings of the National Academy of Sciences of the United States of America 86:9986–9990

Gyllenstein UB, Lashkary D, Erlich HA (1990) Allelic diversification at the class II DQB locus of the mammalian major histocompatibility complex. Proceedings of the National Academy of Sciences of the United States of America 87:1835–1839

Hein J (2004) Pedigrees for all humanity. Nature 431:518–519

Holden C (2001) Oldest human DNA reveals aussie oddity. Science 291:230–231

Horai S et al (1995) Recent african origin of modern humans revealed by complete sequences of hominoid mitochondrial DNAs. Proceedings of the National Academy of Sciences of the United States of America 92:532–536

Ingman M et al (2000) Mitochondrial genome variation and the origin of modern humans. Nature 408:708–713

Jorde LB et al (1997) Microsatellite diversity and the demographic history of modern humans. Proceedings of the National Academy of Sciences of the United States of America 94:3100–3103

Kaessmann H et al. (1999) DNA sequence variation in a non-coding region of low recombination in the human X chromosome. Nat Genet 22:78–81

Krings M, Stone A, Schmitz R-W, Krainitzki H, Stoneking M, Pääbo S (1997) Neandertal DNA sequences and the origin of modern humans. Cell 90:19–30

Krings M, Capelli C, Tschentscher F, Geisert H, Meyer S, von Haeseler A, Grossschmidt K, Possnert G, Paunovic M, Pääbo S (2000) A view of Neandertal genetic diversity. Nat Genet 26:144–146

Lai CSL, Fisher SE, Hurst JA, Levy ER, Hodgson S, Fox M, Jeremiah S, Povey S, Jamison DC, Green ED, Vargha-Khadem F, Monaco AP (2000) The SPCH1 region on human 7q31: genomic characterization of the critical interval and localization of translocations associated with speech and language disorder. Am J Hum Genet 67:357–368

Lai CSL, Fisher SE, Hurst JA, Vargha-Khadem F, Monaco AP (2001) A forkhead-domain gene is mutated in a severe speech and language disorder. Nature 413:519–523

Lai CSL, Gerrelli D, Monaco AP, Fisher SE, Copp AJ (2003) FOXP2 expression during brain development coincides with adult sites of pathology in a severe speech and language disorder. Brain 126:2455–2462

Lawlor DA et al. (1988) HLA-A and B polimorphisms predate the divergence of humans and chimpanzees. Nature 355:268–271

Li W-H (1997) Molecular evolution. Sinauer, Sunderland, MA

Lieberman DE, McCarthy RC (1999) The ontogeny of cranial base angulation in humans and chimpanzees and its implications for reconstructing pharyngeal dimensions. J Hum Evol 36:487–517

Lieberman DE, McCarthy RC, Hiiemae KM, Palmer JB (2001) Ontogeny of postnatal hyoid and larynx descent in humans. Arch Oral Biol 46:117–128

Lieberman P, Crelin ES, Klatt DH (1972) Phonetic ability and related anatomy of the newborn, adult human, Neanderthal man, and the chimpazee. Am Anthropol 74:287–307

Liégeois F, Baldeweg T, Connelly A, Gadian DG, Mishkin M, Vargha-Khadem F (2003) Language fMRI abnormalities associated with FOXP2 gene mutation. Nat. Neurosci 6:1230–1237

Lowenstein JM (1986) Outplaying nature (counterpoints in science). Pacific Discovery 39:37–39

Marsh SGE, Bodmer JG (1991) HLA class II nucleotide sequences. Immunogenetics 33:321–334

McDevitt H (1995) Evolution of Mhc class II allelic diversity. Immunology Rev 143:113–122

McHenry HM, Coffing K (2000) *Australopithecus* to *Homo*: transformations in body and mind. Annu Rev Anthropol 29:125–146

Nordborg M (1998) On the probability of Neandertal ancestry. Am J Hum Genet 63:1237–1240

O'Huigin C, Bontrop RE, Klein J (1993) Nonhuman primate MHC-DRB sequences: A compilation. Immunogenetics 38:165–183

Ovchinnikov IV, Götherström A, Romanova GP, Kharitonov VM, Lidén K, Goodwin W (2000) Molecular análisis of Neandertal DNA from the norteen Caucasus. Nature 404:490–493

Relethford JH (2001) Absence of regional affinities of Neandertal DNA UIT living humans does not reject multiregional evolution. Am J Phys Anthropol 115:95–98

Rohde DT, Olson S, Chang JT (2004) Modelling the recent common ancestry of all living humans. Nature 431:562–

Rogers AR, Jorde LB (1995) Genetic evidence on modern human origins. Hum Bio 67:1–36

Ruvolo M (1997) Mollecular philogeny of the hominoids: Inferences from multiple independent DNA sequence data sets. Mol Biol Evol 14:248–265

Ruvolo M et al. (1993) Mitochondrial CO II sequences and modern human origins. Mol Biol Evol 10:1115–1135

Sarich W, Wilson AC (1967a) Rates of Albumin evolution in primates. Proceeding of the National Academy of Sciences 58:142–148

Sarich W, Wilson AC (1967b) Immunological time scale for hominid evolution. Science 158:1200–1203

Schmitz RW, Serre D, Bonani G, Feine S, Hillgruber F, Krainitzki H, Pavo S, Smith FH (2002) The Neandertal type site revisited: interdisciplinary investigations of skeletal remains from the Neander Valley, Germany. Proceedings of the National Academy of Science USA 99: 13342–13347

Serre D, LanganeyA, Chech M, Teschler-Nicola M, Paunovic M, Mennecier P, Hofreiter M, Possnert G, Pavo S (2004) No evidence of Neandertal mtDNA contribution to early modern humans. PloS Biol 2:313–317

Shu W, Yang H, Zhang L, Lu MM, Morrisey EE (2001) Characterization of a new subfamily of winged-helix/forkhead (Fox) genes that are expressed in the lung and act as transcriptional repressors. J Biol Chem 276:488–497

Smouse PE, Li WH (1987) Likelihood analysis of mitochondrial restriction clavage patterns for the human-chimpanzee-gorilla trichotomy. Evolution 41:1162–1176

Steadman HH, Kozyak BW, Nelson A, Thesier DM, Su L, Low DW, Bridges CR, Shrager JB, Minugh-Purvis N, Mitchell MA (2004) Myosin gene mutation correlates with anatomical changes in the human lineage. Nature 428:415–428

Stoneking M et al. (1990) Geographic variation of human mitochondrial DNA from Papua New Guinea. Genetics 124:717–733

Stringer CB, Andrews P (1988) Genetic and fossil evidence for the origin of modern humans. Science 239:1263–1268

Stringer CB (1992) Reconstructing recent human evolution. Philos Trans R Soc 337:217–224

Templeton AR (2002) Out of Africa again and again. Nature 416:45–51

Thorne AG, Wolpoff MH (1992) The multiregional evolution of humans. Sci Am 266:76–83

Tobias P (1986) Delineation and dating of some major phases in hominidization and hominization since the middle miocene. S Afr J Sci 82:92–94

Tobias P (1991) Relationship between apes and humans. In: Sahni A, Gaur R (eds) Perspectives in human evolution, Renaissance, Delhi, pp 1–19

Vargha-Khadem F, Watkins KE, Alcock KJ, Fletcher P, Passingham R (1995) Praxic and nonverbal cognitive deficits in a large family with a genetically transmitted speech and language disorder. Proceedings of the National Academy of Science USA 92:930–933

Vargha-Khadem F, Watkins KE, Price CJ, Ashburner J, Alcock KJ, Connelly A, Frackowiak RSJ, Friston KJ, Pembrey ME, Mishkin M, Gadian DG, Passingham RE (1998) Neural basis of an inherited speech and language disorder. Proceedings of the National Academy of Science USA 95:12695–12700

Vigilant L et al. (1991) African populations and the evolution of human mitochondrial DNA. Science 253:1503–1507

Waddle DM (1994) Matrix correlation tests support a single origin for modern humans. Nature 368:452–454

Watkins KE, Vargha-Khadem F, Ashburner J, Passingham RE, Connelly A, Friston KJ, Frackowian RSJ, Mishkin M, Gadian DG (2002) MRI analysis of an inherited speech and language disorder: structural brain abnormalities. Brain 125:465–478

Wolpoff MH, Hawks J, Caspari R (2000) Multiregional, ot multiple origins. Am J Phys Anthropol 112:129–136
Wolpoff MH (1988) Neanderthals: not so fast. Science 282:1991
Wood B, Collard M (1999) The human genus. Science 284:65–71
Zimmer C (1999) New date for the dawn of dream time. Science 284:1243–1246

CHAPTER 2

GENETICS AND NEUROSCIENCE

Some examples of their recent convergence and of the continuing nature–nurture controversy, with emphasis on sleep physiology

CLAUDE DEBRU

Chair of Philosophy of Science, Ecole Normale Supérieure, 45 rue d'Ulm, Paris, France
claude.debru@ens.fr

Abstract: Neurogenetics is a more recent research field than neuroscience. It is characterized by the introduction of genetical techniques and techniques of molecular biology in many fields like the study of learning and memory, language, pathology etc. The field of neurogenetics merges increasingly with biological psychology. In this paper, early studies by Eric Kandel showing the dependency of long term memory on a switch of genes which enhance the learned response, as well as later studies of learning using molecular biology techniques are discussed. Sleep physiology and pathology is discussed with special emphasis on Michel Jouvet's reprogramming theory of paradoxical sleep, which insists on the potential regulatory function of paradoxical sleep on the interaction between genetic and epigenetic features, providing a possible answer to the nature–nurture controversy

Neuroscience as an integrated interdisciplinary field goes back to the late sixties and early seventies, when the MIT study programme bearing this name started. Neurogenetics is a more recent field, which really began when the study of biochemical mechanisms in the brain was mature enough to allow genetic techniques (inbred, mutant, transgenic or knock-out mice) to be fruitfully introduced, thus helping to understand the mechanisms of various diseases as well as of normal brain function relying on an increasing number of transmitters and receptors. The genetics of brain development may be mentioned also as a most important field of research, which developed recently (this field should be more appropriately considered in a chapter on genetics and development). However, recent scientific advances in neurogenetics should not be separated from other formerly established research fields like the physiology of behavior or psychology, which provided much material for pre-existing ideological discussions like the nature–nurture controversy, still going on in a very fierce way.

As an example of the involvement of all these various scientific, philosophical and ideological aspects in contemporary research, I will deal with a particular domain,

A. Fagot-Largeault et al. (eds.), The Influence of Genetics on Contemporary Thinking, 25–32.

sleep and dream physiology, which seems particularly suitable to demonstrate how facts and speculations coexist and interact in science, but also to demonstrate how the genetic approach may be useful to solve problems in physiology and pathology. However, before entering the subject of sleep physiology (largely shaped by Michel Jouvet, William Dement and their colleagues in the last fifty years), it may be useful to make some comments on other aspects of the involvement of genetics in earlier physiology and psychology, as well as in contemporary neuroscience, including its behavioral and cognitive dimensions.

Many fields (virtually every field) of contemporary neuroscience are shaped by genetic research, sometimes in a less expected and obvious manner than others. Memory and learning, language, neurological diseases, circadian rhythms and sleep patterns are some examples of the genetic components increasingly found in the workings of the brain. A most revealing and unexpected example of the role of genes in physiology is found in the studies on memory and learning carried on by Eric Kandel during the last fifty years on the marine mollusk *Aplysia californica*. Indeed, the idea that learning (or at least part of its functioning) takes place thanks to the switch of a genetic mechanism of transcription and translation may seem at first rather counterintuitive. In a recent account of his work, Kandel has beautifully shown the way his research progressed from the study of synaptic plasticity and its role in short-term memory to the discovery of the role of genetic events in long-term memory. This distinction between short-term and long-term effects in memory goes back (at least) to the end of the 19th century.

> At the end of the 19th century, Hermann Ebbinghaus and William James independently began the experimental examination of human memory with a series of groundbreaking observations. A century of investigation into the nature of memory has supported many of their central contentions, including the observation that human memory is not a unitary process of information storage but rather consists of multiple processes, each with a different time course and behavioral role. One distinction that emerges with particular clarity is that memory has distinct temporal phases.
>
> (Pittinger and Kandel, 1998, 91)

The initial phase results in labile, easily disrupted storage; a later phase results in consolidated memory. These concepts were later verified in studies like the ones performed by Kandel on synaptic plasticity (which was already recognized by Ramon y Cajal as a process of information storage). Working with *Aplysia*, and later with mice, Kandel did very revealing observations.

> In *Aplysia*, we have examined sensitization of the gill-withdrawal reflex, a simple, non-associative learning paradigm. When the animal is touched on the siphon, it retracts its gill. A single electric shock to the tail enhances the withdrawal response for a period of several minutes. Protein synthesis inhibitors do not affect this short-term sensitization. In contrast, a series of four or five shocks to the tail produces sensitization lasting for days. Protein synthesis inhibitors delivered at the time of the tail shocks prevent formation of this long-term memory. Sensitization of the gill withdrawal reflex thus exhibits the protein synthesis-independent short phase and the protein synthesis-dependent late phase that characterize more complex forms of memory in the mammalian brain.
>
> (ibid., 92)

The fact that consolidated memory depends on protein synthesis can be only explained by a switch of a genetic transcription-translation mechanism. The biochemical mechanisms of this induction include an up regulation of genes by an increase of cAMP and by other events in the nucleus of the neuron leading to long-term facilitation, it means to long-term enhancement of the reflex response. In mice, similar events, factors and mechanisms seem to take place in long-term potentiation (LTP).

In the brain of mammals like rabbits, studies of LTP in the hippocampus gave support to the view that LTP plays a major role for the encoding of memories (Davis and Laroche, 1998, 98). From the original discovery of T.V.P. Bliss and T. Lomo in 1973 onwards, these concepts based on Hebb's and Konorski's theories on synaptic memory storage were applied to distributed networks, not only to simple circuits, and gained increasing attention. Various forms and aspects of LTP were mentioned as proving the function of LTP in memory storage. They included associative learning (learning by activating a subset of converging inputs rather than the whole set of them), cooperativity (threshold of input, and recruitment of a greater number of fibres), specificity (in the sense that not only primarily activated synapses are subject to LTP, but that synapses in the vicinity of these may also exhibit potentiation). Synaptic mechanisms include the well-known activation of the NMDA receptor. The study of synaptic plasticity in terms of biochemical and morphological changes underwent recently further changes thanks to the introduction of the techniques of molecular biology. According to Serge Laroche,

> perhaps one of the most important discoveries has been that genes do not only act as the backbone or the house-keeper for maintaining the integrity of the cell but can be rapidly and functionally activated or deactivated in response to some form of neuronal activity. The tools available to hand now are those in which we are able to measure the activity of genes, to delete a specific gene target with genetically engineered mutant mice, to prevent the gene manufacturing new proteins with antisense oligonucleotides. With these we are able to access more defined mechanisms involved in synaptic plasticity to understand what are the potential events that lead to the encoding and storage of memories.
>
> (ibid., 100)

Molecular biology techniques revealing gene expression could help recently to go beyond single synapses and to study "transsynaptic plasticity".

> The fact that genes can be functionally activated in response to neural activation ... would not have been conceived of a decade or two ago, let alone the notion that they can be naturally regulated during learning.
>
> (ibid., 104)

Learning and memory are just a very revealing case of the power of molecular biology techniques to unravel unexpected mechanisms of gene expression in a number of neurobiological and psychobiological phenomena. Neurogenetics developed in recent years as a new specialty, with emphasis on the study of synaptic mechanisms by molecular biology techniques (mutants, transgenic animals).

Biological psychology provided examples of the wide distribution of learning abilities in animals. *Drosophila* is able to learn as well as *Aplysia* and can be trained.

The study of learning in *Drosophila* started in 1974, when William Quinn was able to condition these flies with electric shocks and odors. Then mutant strains unable to perform the discrimination tasks involved in learning were isolated. Several stages in the learning mechanisms of *Drosophila* were identified more recently. Genetics seems to confirm the existence of these separate stages, since impairment of specific genes or of groups of genes results in specific deficiencies for each stage. (Rosenzweig et al., 2005, 575). However, the idea that genetic factors sufficiently explain the kind of learning that a species is able to perform was questioned. For instance, some birds do not recognize the color of their eggs but recognize the chicks within three days after the hatch. The current interpretation of this learning selectivity does not any more rely on genetic constraints but on evolutionary and ecological mechanisms of selective pressure (ibid., 545).

These results and debates show that the field of neurogenetics gained recently a more secure scientific basis. However, in the 1960s, at the time of the major developments in molecular biology, when the metaphors of genetic information, genetic code and genetic programming became common parlance among scientists, brain scientists or at least those of them who were more theoretically oriented did try to make real use of these ideas in their own interpretations, questions and theories. Sleep physiology provides a good example of the early introduction of the vocabulary and concepts of genetics, information theory, and molecular biology in neuroscience. In 1967, Edmubnd Dewan, a scientist of the US Air Force, proposed a new and original hypothesis about the function of rapid eye movement sleep (or paradoxical sleep in Michel Jouvet's terminology). In 1953, Eugene Aserinsky working in Nathaniel Kleitman's sleep research laboratory discovered the occurrence of rapid eye movements phases during sleep in newborns. This discovery was the starting point of major developments in neurophysiology because the phases became soon associated with the occurrence of dreaming, especially in the work of William Dement. In 1959, Michel Jouvet discovered independently the occurrence of particular sleep phases in decorticated cat, characterized by a low voltage fast electrical activity and a total disappearance of muscular tonic activity. He coined the phrase 'paradoxical phase' to describe the coexistence of intense internal brain activity and almost total absence of peripheral activity. This discovery gave even more prominence to the phenomenon of paradoxical or rem sleep, because the question of its physiological significance became more pressing. Which special kind of process is going on in the brain during paradoxical or rem sleep? One tentative answer to this question was given by Edmund Dewan (1970, 296) in 1967, when he proposed that the sleeping brain, during the rem phase, carries on a programming or reprogramming activity. In his view, the permanent adaptation of the brain to new tasks implies that the brain is able to activate certain programmes and put other ones in memory. This theoretical view proposed by a computer scientist relies also on another assumption, that the brain contains programmes (mainly behavioral ones) which are encoded in specialized networks of neurons. In the brain, these programmes would need to be reactivated, reread or even rewritten repeatedly. Such a reprogramming was supposed to occur when the brain is inactive, it means 'off line'. According to Dewan, the brain does own such a reprogramming function, which is activated during rem sleep.

Michel Jouvet did not consider at first this hypothesis. He had other views as a physiologist dealing with states of sleep and wakefulness. Paradoxical sleep, associated with dreaming, may be viewed as a state of arousal within sleep. In a review devoted to biochemical data on sleep, Michel Jouvet first proposed to consider paradoxical sleep as a 'genotypic arousal'. The phasic, repeated stimulation of the cortex is the most striking phenomenon of paradoxical sleep. Its function could be to ensure and to keep the specificity of connections between interneurones during and after brain maturation. These networks of late connecting interneurones would provide the basis for coding innate behaviors. According to Jouvet's proposal in 1972, the function of paradoxical sleep during brain maturation would thus be "genotypic coding" (Jouvet, 1972, 270). In adult life, paradoxical sleep would maintain the specificity of connection of these interneurons, and control the possible interference of epigenetic and genetic events. These speculations lead Jouvet to ask a particular question: does a genetic programming of the brain occur during paradoxical sleep? Asking this question was easier and perhaps even necessary after the discovery of the behavioral component of paradoxical sleep. The animal (cat) which underwent a surgery destroying the small structure of the midbrain named locus coeruleus loses the motor inhibition which is a character of paradoxical sleep. As a consequence, he shows, during paradoxical sleep episodes, sequences of well-defined behaviors which are quite well observed and related to standard patterns, like watching a prey, fearing, etc. If the genetic interpretation of these behaviors holds true,

> if we admit that genetic factors are partly at least responsible for the overall expression of behavior, we should be able to to answer the following question: are the templates for 'innate releasing stimuli' and the neural systems for 'fixed motor patterns' genetically programmed, structurally or functionally, once for all, when the maturation of the nervous system is achieved? In other words, will the subtle organization of connections between those synapses responsible for innate behavior or for inherited differences of behavior between individuals remain unaltered throughout the life span? Moreover, how can similar behavior patterns appear in squirrels that have been raised in totally different surroundings and have been submitted to different historical (epigenetic) stimuli? Numerous experiments have demonstrated conclusively that the environment may indeed alter the nervous system both functionally and structurally.
>
> (Jouvet, 1978, 245–246)

In order to answer these questions, Michel Jouvet devised a theoretical model of periodic genetic programming of the central nervous system, whose basic propositions were established in such a way that paradoxical sleep would be the carrier of this function. Notably enough, protein synthesis was part of the picture, as a precondition of the programming process. It was recognized that "genetic programming"

> is a rather ambiguous term: it is used here to mean a periodic endogenous program which maintains, facilitates or inducts the organized systems of neurons responsible for the innate releasing templates and fixed motor patterns which are used in innate behavior or responsible for inherited typology.
>
> (ibid., 246)

In other words, there would be a special programme, possibly contained in a protein and transduced according to some neuronal coding into specific patterns of firing or spatial distributions of active neurons, which would carry out the "genetic readout" (ibid., 247). Temporal coding was supposed to occur through the sequence of the ponto-geniculo-occipital (PGO) waves which propagate from the generator of PS in the pons to the cortex. This visionary theory was never proved (nor entirely disproved). However, most of its content continues to inspire and influence many speculations regarding the function of paradoxical sleep.

This example of the highly speculative kind of influence of genetics on functional theories in neuroscience should not be underestimated. True, what is genetically reprogrammed during paradoxical sleep remains to be established. It is no more believed that instinctive behaviors need such a reprogramming, since for instance suppression of paradoxical sleep in rats does not impair maternal behavior towards the newborns. A more recent version of the theory states that paradoxical sleep maintains phenotypic variance among individuals. As a matter of fact, suppression of paradoxical sleep in different stocks of inbred mice has effects on their performances in memory tasks. Fast learning mice are affected by paradoxical sleep deprivation, which diminishes their performance, while slow learners are unaffected. However, paradoxical sleep deprivation does not affect the basic mechanisms of the learning process itself. The relationship between paradoxical sleep and memory consolidation remains controversial. The extent of genetic determination of brain functions may be seen in sleep patterns and circadian rhythms, which are quite different in different inbred mice, but show an intermediate shape in hybrids. Not only sleep patterns, but also the number of neurones in a single hypnogenic structure may vary in different inbred mice, as shown by Jean-Louis Valatx (Valtax et al., 1982). The extent of genetic variation in biological structures is always a striking, but not an unexpected phenomenon. Contemporary genetic analysis provides new data and solutions to a particular problem of sleep pathology, the pathophysiological mechanism of narcolepsy-cataplexy (Gélineau disease), thus throwing important additional light on sleep and waking mechanisms. Working in William Dement's Sleep Research Center at Stanford University, Emmanuel Mignot studied the family of narcoleptic dogs raised by William Dement and was able to unravel the pathophysiological mechanisms of the disease by studying at the same time the genetics (chromosomal mutation) and the biochemistry (peptides and receptors) of these narcoleptic dogs. Emmanuel Mignot's (2004) results gave arguments for the view that sleep and wakefulness are primarily controlled by hypothalamic structures.

It is increasingly recognized that genetic mechanisms influence cognitive processes. Pathology provides many other examples of this. However, many specialists do agree that this statement does not imply a paramount causal power of genetics on the various phenomena studied by cognitive neuroscience. The keyword here is "plasticity". In spite of its very diffuse meaning and of its very different levels of application, plasticity remains a useful term once it is understood that the fact to which it alludes, i.e. that a modification of form and function, once imprinted, remains as a more or less permanent property, is based on the ability of genes to respond readily to stimuli in their cellular environment, thus enhancing already established responses. In a way, the fact that plasticity rests basically on genetic mechanisms shows how deeply the apparent opposition between plasticity and genetic determination remains caught within the paramount power of genetics. The genes do provide the basic conditions of cell functions. However, synaptic degeneration, not only plasticity, should be also involved in this discussion. This idea was perfectly understood by Jean-Pierre Changeux in the early seventies, when he proposed the notion of 'genetic envelope', which can be readily applied to contemporary discussions in neuroscience and genetics. Genes do not determine entirely every phenotypic character. They provide a background of possible paths of development, which are characterized by the observable disappearance of existing connections and by the enhancement of other ones. Molecular biology tools do reveal the extent of these phenomena of plasticity and degeneration. Plasticity remains an important property of the central nervous system until the end of the life. It is true that genetic and epigenetic influences act both on the system in different ways. However, the lesson of recent biological work is double: epigenetic influence may depend on genetic mechanisms for its expression; specificity of genetic action remains debatable. Consequently, a dialogue between genetic and epigenetic action should take place in the brain. This dialogue needs some form of regulation and arbitration. Paradoxical sleep could be a good candidate for playing such a regulative role at a very general level. Nature and nurture are not contradictory terms. Twenty five years ago, Michel Jouvet (1980, 343) suggested that many possibilities of interaction between nature and nurture might exist, and that programming mechanisms would increase the range of modifiability of innate behavior in the nervous system. In current speculations, a modified version of the same basic idea could be that reprogramming mechanisms could by their very existence increase the range of modifiability and phenotypic variance among individuals. As mechanisms produced by evolution and natural selection, they could contribute to an increased phenotypic variation.

REFERENCES

Davis S, Laroche S (1998) A molecular biology approach to synaptic plasticity and learning. International symposium: memory, from neurone to cognition, Comptes rendus de l'Académie des sciences, Sciences de la vie/Life sciences, Paris, Sciences de la vie/Life sciences, 321

Dewan EM (1970) The programming (P) hypothesis for REM sleep. In: Hartmann E (ed) Sleep and dreaming, international psychiatry clinics, vol 7, n 2, Little Brown, Boston

Jouvet M (1972) The role of monoamines and acetylcholine containing neurons in the regulation of the sleep-waking cycle. Ergeb. Physiol. 64, 166–307

Jouvet M (1978) Does a genetic programming of the brain occur during paradoxical sleep? In: Buser P and Rovgeul - Buser A (eds), cerebral correlates of conscious experience, INSERM Symposium n° 6, Elsevier/North Holland Biomedical Press, Amsterdam.

Jouvet M (1980) Paradoxical sleep and the nature–nurture controversy In: P.S. McConnell, GJ. Boer, H.J. Romijn, N.E. Van de Poll and M.A. Corner (Eds.) Adaptive capabilities of the nervous System, Progress in Brain Research, vol 53; Elsevier, Amsterdam, 331–46

Mignot E, Fromherz S (2004) Narcolepsy research: past, present and future perspectives. Arch Ital Biol 142: 79–486

Pittenger C, Kandel E (1998) A genetic switch for long-term memory.International symposium: memory, from neurone to cognition, Comptes rendus de l'Académie des sciences, Sciences de la vie/Life sciences, Paris 321

Rosenzweig MR, Breedlove MS, Watson NV (2005) Biological Psychology. An introduction to behavioral and cognitive neuroscience, Sinauer Associates, Sunderland Massachusetts

Valaxt JL, Touret M, Jouvet M (1982) The locus coeruleus: a quantitative and genetic study in mice. Brain Res. 250 (2), 353–357

CHAPTER 3

WHO MADE THE GENETIC CODES, HOW AND BY WHAT?

KOICHIRO MATSUNO

Department of BioEngineering Nagaoka University of Technology, Nagaoka, Japan, kmatsuno@vos.nagaokaut.ac.jp

Abstract: Evolutionary emergence of the genetic codes is a material manifestation of the indexical activity latent in chemical molecules. Descriptive access to the indexical activity of material origin is made possible in first and second person description. The first law of thermodynamics on energy conservation during its transformation, when referred to in second person description, manifests the activity for fulfilling energy conservation from within, while there is no such activity identifiable when accessed in the standard third person description. Focused on material activities accessible exclusively in second person description is the process of measurement internal to the material bodies. Internal measurement distinguishes between the two tenses, namely, present progressive and present perfect tenses. Once internal measurement precipitates a material representation accessible in the present perfect tense, the representation can be referred to even in the present tense in third person description. The genetic codes are the supreme example of material representations that can be precipitated from internal measurement proceeding in the primordial soup of reacting chemical molecules that is accessible intrinsically in second person description in the present progressive tense only

1. INTRODUCTION

Facing the origin of the genetic codes is causally reversed in letting our natural language as the historical latecomer to address the predecessors. If the intended endeavor on deciphering the historical origin under the causally reversed context has any positive implication whatsoever, it must be prerequisite to distinguish what is sequential in time from what is concurrent. One signpost for the directive has been the notion of interaction taken up by Kant as a condition on stipulating the community of things to coexist in a concurrent manner. Interaction as a condition for things to synchronously coexist is certainly the case with action at a distance in Newtonian mechanics. Nonetheless, Kantian-Newtonian interaction has difficulty in accommodating sequential interaction in a manner consistent with synchronous

A. Fagot-Largeault et al. (eds.), The Influence of Genetics on Contemporary Thinking, 33–50.

one except for appealing to the subjective condition of the observer external to those things to be experienced. It is one thing that sequential interaction in time or causality is made at home with the external observer, but quite another whether or not it may be approachable only to the subjective observer sitting outside (Matsuno, 2000; Rosen, 2004).

Both classical and quantum mechanics have been practiced mainly in the tradition that synchronous interaction is overly emphasized in comparison to sequential one as exemplified in the form of an equation of motion. The scheme of equating the successor to the predecessor is as a matter of course upon a metaphysical stipulation of concentrating only on synchronous interactions as dismissing sequential ones. In fact, it would seem almost irresistible to emphasize the significance of synchronous interaction out of proportion once one is determined to commit oneself exclusively to third person description in the present tense.

The essence of third person description in the present tense is in the methodological stipulation on insisting the proclaimed legitimacy of the attempted statement either temporally at any present moment or atemporally without referring to any temporal dimension. If it were conceivable to have a set of fundamental predicates that may remain valid all through the historical or evolutionary development of the material world, descriptive analysis of the evolution upon third person description in the present tense thus framed would stand the test of time. Otherwise, it would be required to have a descriptive scheme other than those restricted solely to third person description in the present tense. The issue of the origin of the genetic codes may certainly be the case in point. Focused upon in this regard will be the relevance of both the present progressive and present perfect tenses in other than third person description in addressing the origin of the genetic codes.

Of course, it goes without saying that the origin of the genetic codes must have been a consequence of evolutionary movement that would have proceeded on our primitive Earth. What is more, any movement on the spot is in the present progressive mode. This reflection now invites us to figure out how the present progressive tense could be approached as facing evolutionary processes in the material world. At issue will be how the present progressive tense could eventually be addressed in the present tense, since third person description in the present tense would become inevitable in any descriptive endeavor even including this article attempted by the present author.

2. PUNCTUATION OF THE PRESENT PROGRESSIVE TENSE

Material dynamics is nothing other than about material movements in progress. And the representative case of material dynamics comes from quantum phenomena. An empirical basis of quantum dynamics resides in the activity in the present progressive tense demarcated clearly by the occurrence of its completion referred to in the present perfect tense. When a black body emits and absorbs light waves, Max Planck observed that light emission in progress is always punctuated by its completion. The discontinuous distinction between light emission in progress and its completion is legitimately materialistic in its origin. A photon as a propagating wave is in the present progressive mode, while a photon as a particle serves as the container of the propagating wave whose progressive movement inside has been perfected at the contour of the container (Matsuno, 2002b, 2003).

The distinction between the present progressive and the present perfect tenses is made from within, and the associated measurement is internal (Matsuno, 1989). The activity of measuring the contour at which the movement from the inside has been perfected is indexical internally and addressable exclusively in second person description because of the indexical nature. What is unique to internal measurement on the part of an energy quantum in general and a photon in particular is the natural transference from the present progressive to the present perfect tense at the contour of the quantum. An energy quantum after Planck is a material embodiment of both movement in progress and movement perfected. The distinction between the inside and the outside of a quantum is to be accomplished through internal measurement proceeding there. Internal measurement viewed from the perspective of quantum dynamics is agential in the act of transferring the present progressive tense into the present perfect one and serves as the material agency bridging the present progressive and the present perfect tenses. Quantum dynamics thus empirically provides a material support addressed in second person description in the present progressive tense, the latter of which is necessarily punctuated by the present perfect tense from time to time from within. That is agential from within, instead of from without.

At the same time, mechanistic causation remains invincible insofar as the externalist stance guaranteeing the certitude and integrity of the descriptive object out there in the third person status is sanctioned. The causation toward each individual is definite and proceeds in a completely consistent manner with the rest constituting the whole object. Complete definiteness of mechanistic causation dispenses with even the notion of context and contextualization since the definiteness has already been guaranteed for every individual constituting the whole context of whatever kind. Mechanistic causation for individualization is already implicit with complete contextualization in third person description. On the other hand, actual causation for contextualization goes beyond simply being mechanistic when the completeness of individualization is jeopardized for whatever reasons (Küppers, 1995). Final causation will enter for contextualization when there remains some indefiniteness in materializing each individual cause. The lack of complete individualization of causes will arise when each individual cause comes to influence and communicate with others. Final cause for the context then available is an inevitable participant in causation when there arises a conflict between the communication for causation and the caused movement. Unless the caused movement simultaneously turns out causation to others, the communication of causation not synchronized with the caused movement becomes real. The communication of causation is mechanistic in leaving the caused movement behind, but final at the same time in acting toward conformity with the context. Communicating of causation is mechanistic in its individualization, while being final in its contextualization.

The distinction between contextual and mechanistic dynamics will become obvious when one distinguishes between second and third person descriptions to be employed. Dynamics addressable in third person description has to be mechanistic, since third person description admits the agential capacity of neither the speaker or writer of the utterance in which it occurs nor the one to which that utterance is addressed. The agential capacity influencing the content of the utterance in third person description is always sought outside the utterance. In contrast, dynamics addressed in second person description can be contextual since second person description refers to and appreciates the agency of a thing addressed in the utterance in which it occurs. The thing addressed in second person description can maintain room of influencing and modifying further the context of the utterance that has already been done by the speaker or writer. What is more, this distinction between contextual and mechanistic dynamics is not merely a matter of linguistic artifact. The origin of the object addressable in second person description is even quantum dynamical empirically.

3. CONTEXTUAL DYNAMICS

One more representative case of contextual dynamics comes from thermodynamics (Matsuno, 2001). What is unique to thermodynamics is that it introduces macroscopic variables such as volume, pressure, temperature and entropy without detailing their atomistic makeup at the outset. These macroscopic variables are about the context in which the underlying microscopic elements, whatever they may be, are eventually situated. The contextual dynamics specifying the values of the macroscopic variables is constantly operative there. Even if the fundamental dynamics of microscopic elements is left unspecified, the contextual specification is to proceed. A molecule in the gas is subject to the temperature of the gas while at the same time the molecule is part of the gas substantiating the same temperature. Thus, any contextual element constituting the context comes to materialize and share the same contextual specification. Thermodynamics is unique in emphasizing the priority of contextual specification over elementary specification of each constituent element. Although mechanics is a theoretical enterprise equating elementary specification of an imposed character literally to contextual one in a crisp manner, thermodynamics is quite different in allowing an under-complete elementary specification whether or not it is of an imposed character. The present contextual specification now provides the interplay between the two of the contextual and the elementary dynamics with a possibility of influencing each other in both ways, namely, from the elementary to the contextual and vice versa.

One attempt for relating the elementary to the contextual dynamics is through a statistics of mechanics over an ensemble of elementary specifications. Statistical mechanics is grounded upon the premise that an ensemble of elementary specifications could be a substitute for the interplay between the two specifications, the contextual and the elementary ones. A justification of the ensemble of elementary specifications came from Boltzmann's Stosszahl Ansatz or hypothesis of a molecular chaos stating that molecules in the gas are going to lose their memory of the past collisions with the others soon except for the latest ones. Those molecules in the gas thus come to have almost no correlation with the others or to move almost randomly with each other. This is equivalent to saying that the context which Boltzmann introduced is the one under which every contextual element moves almost randomly with each other. More specifically, one particular quantitative figure characterizing the Boltzmann's context is called temperature. To be sure, Boltzmann's context is found ubiquitous in physics. Nonetheless, it is no more than a heuristic candidate for fulfilling the role of contextual specification operating in thermodynamics. There certainly is another candidate for meeting the similar requirement of contextual specification. That is from quantum dynamics.

4. CONTEXTUAL SPECIFICATION OF A QUANTUM

The context or the contour specifying a quantum of whatever kind is naturalized of itself. Naturalization means restricting theoretical artifacts to a bare minimum to the extent that can be tolerated while making access to empirical reality as closely as possible. The context of material dynamics is thus constructed in a bottom-up manner, while the boundary condition addressing the context theoretically is imposed in a top-down manner. Naturalization of the context in biology is in the effort of minimizing the top-down influence of imposed character in describing biological organization (Matsuno, 2000).

Atoms and molecules constituting a biological organism are placed within the material context of an extremely specific configuration. Such a specificity of the material context makes biological organization unique compared to nonliving physical organization of atoms and molecules. Needless to say, physics has its own rich history on explicating the nature of whatever material contexts available there. One example of an extreme significance is the material context discovered by Max Planck.

First of all, context as a limiting modifier of the contextual elements, if empirically available, must remain robust to a reasonable extent. Otherwise there would be no possibility for denoting it as such in the empirical domain. Empirical confirmation of the occurrence of such a robust context including a set of macroscopic variables in thermodynamics comes from examining the empirical record of the events of interest. The record is about the events already registered in the present perfect tense, while the ones right in the making are in the present progressive tense. The robust record rests upon the transference of events in the present progressive tense to those registered in the present perfect one. In fact, quantum dynamics grounds itself upon the existence of such a robust record.

When Planck introduced the notion of a quantum for the first time, the relevant empirical fact referred to was that a light-wave emission from and absorption to a black body in thermal equilibrium with its surroundings are punctuated in a discrete manner. The discreteness is associated with the empirical observation that light-wave emission in progress comes to shortly be punctuated by the emission completed, and light-wave absorption in progress similarly comes to be punctuated by the absorption done. There is no indefinite prolongation of light-wave emission and absorption over to an infinite duration in a continuous manner. The punctuated light-wave referred to as a light quantum or a photon carries with itself the context within which continuous light-wave is encapsulated in a coherent manner. In particular, Schrödinger identified that the coherent nature of the encapsulation is due to the occurrence of a standing wave as a coherent superposition of both the retarded and the advanced waves of material origin.

The context discovered by Planck, or Planck's context, is thus the one for those contextual elements moving almost coherently with each other. Planck's context is just a polar opposite to Boltzmann's, in the latter of which the contextual elements are taken to move almost randomly in an incoherent fashion with each other. However, the relationship between Planck's context and Boltzmann's is not mutually exclusive. Planck's context is more fundamental and more inclusive in that any material element of whatever sort is a quantum after Planck. In contrast, Boltzmann's context is subject to Planck's contexts embedded in it. At the same time, Planck's context is also subject to influences coming from the outside, because it always presumes the action of making a sharp distinction between the present progressive and present perfect tenses by both itself and others external to the context itself. What is responsible for generating the context is the robust interplay between the inside and the outside. The present interplay can now furnish a Boltzmann's context as a source matrix of the measuring agencies toward each Planck's context residing in its inside with the capacity of modifying the latter context in time internally. Occurrence of a Boltzmann's context is in fact an empirical testimony to the observation that the constituent quanta or Planck's contexts are measuring each other internally, that is to say, involved in internal measurement altogether (Matsuno, 2001).

In particular, the contrast between Planck's context and Boltzmann's will become more transparent once the nature of internal measurement involved is focused. Although Boltzmann's context rests upon the stipulation that each quantum loses the memory of the past measurements of the others shortly, Planck's context is about the persistent memory of the measurement internal to each quantum while distinguishing the movement in the present progressive mode from the one in the present perfect. Planck's context is for long memory of internal measurement, while Boltzmann's is for short memory.

5. QUANTUM DYNAMICS IN SECOND PERSON DESCRIPTIONS

Cells and tissues met in biological organisms are unquestionably quantum mechanical in their makeup. Nonetheless, the deciphering of their functions has mainly been practiced in terms of the language of cellular and molecular biology instead of that of quantum dynamics. This observation alone seems to suggest that there might be a possibility for exploring quantum dynamics further in order to accommodate to itself an essence of what cellular and molecular biology would imply. Take, for instance, a homeostatic activity of the biological cell. Although it may look tempting to associate the homeostatic capability with the fundamental nature of the boundary conditions that come into being to underwrite the existence of the cell itself, this perspective necessarily tends to beg further questions down the road. This reduction of one problem at a certain level into another one at a more fundamental level cannot however proceed in an infinitely regressive manner. The bottom line is quantum dynamics. It is required to seek the functional origin of homeostatic activities within quantum dynamics, otherwise our endeavor for clarifying material underpinning of biological activities would turn out futile in the end. At issue is the relationship between quantum mechanics and its boundary conditions (Pattee, 1982).

Suggestive to the functional relationship between quantum mechanics and its boundary conditions is the quantum coherence that has traditionally been referred to as the wave-particle duality. That continuous wave-like characteristic is bounded by a particle-like rigid boundary is common to any material element according to quantum dynamics. The quantum coherence addresses the robustness of the wave-like coherent characteristic. One descriptive means of coping with the quantum coherence is to make an appeal to Schrödinger's equation of the wavefunction subjected to given boundary conditions. If one is sure about the robustness of the chosen boundary conditions out there, the resulting wavefunction would definitely meet the requirement for serving as the material carrier of the quantum coherence. Descriptively, this association of the quantum coherence with the wavefunction is practiced in the third person description because the robustness is taken as an authentic object placed and to be found out there in the third person status. Although it is certainly legitimate in its own light, the robustness of quantum mechanics in third person descriptions, however, does not take up the issue of how the robustness would come into being and develop. This incompetence in facing the issue of dynamic robustness is rooted in the stipulation of sticking to third person descriptions, instead of to quantum dynamics per se. Our effort for addressing dynamic robustness met with the biological cells and tissues may benefit from quantum dynamics addressable in other than third person descriptions.

An energy quantum appearing in whatever material level is a form of confining the interactions operating among the constituent elements almost completely, though it is still allowed to interact with other quanta outside through the interactions leaking out of each confinement. In fact, the confinement cannot literally be complete. For instance, a proton is certainly an energy quantum that confines the interactions operating among the constituent quarks, while it can also serve as a constituent element for making an atomic nucleus. A Cooper pair of electrons having anti-parallel spins constitutes an energy quantum whose macroscopic condensation gives rise to superconductivity in metals.

What is specific to each energy quantum as an almost complete confinement of interactions is its robustness against external disturbances. This has been an established empirical fact thanks to Max Planck. The robust confinement of interactions in motion assumes that each party of an arbitrary interacting pair of the confined constituent elements comes to internally measure the other party in the second person status. The process of confinement is actually in the progressive mode, and it is intrinsically prior to being frozen in the completed record registered in the present perfect tense. Movement in the progressive mode fundamentally differs from the recorded movement in the present perfect tense because the temporality addressing the movement is different. The record registered in the present perfect tense in third person description inevitably includes in itself the component of being perfected that is not found within the movement right in progress. Direct access to interactions in quantum dynamics is attempted in second person descriptions since interactions intrinsically assume reciprocity of activities referred to in second person descriptions. Every energy quantum conceived in quantum dynamics is upon the robust confinement of interactions or internal measurement, which is descriptively accessible in the second person status.

In particular, an energy quantum as the robust confinement of internal measurement now serves as a unit of further internal measurement precipitating both the first and the second laws of thermodynamics accessible in the present perfect tense in third person description. This implies that the robust confinement of internal measurement differs from a material unit identified in the completed record registered in the present perfect tense. The robust confinement refers to the confined pattern of interactions in motion, but by no means the completed record. The confinement is in fact a form of memory since it is nothing other than a pattern of interactions in the progressive mode. Interactions in motion cannot survive without their predecessors. Memory in motion cannot be accessible in third person description since it cannot be set as an invariant object in the third person status. If it is approachable descriptively at all, the confined pattern of internal measurement has to be in the second person status in focusing upon the contrast between the confining and the confined. Memory in motion thus makes the occurrence of second person descriptions inevitable within the framework of quantum dynamics.

If some form of internal measurement happens to be confined, the robustness of such a confinement would guarantee its occurrence as a descriptive object exclusively in the second person status. Quantum dynamics empirically confirms the robust confinement of internal measurement in the form of an energy quantum. Second person description enables us to perceive each energy quantum met in quantum dynamics as the robust confinement of internal measurement or as memory in motion.

The robust confinement of internal measurement or interactions in fact assumes the process of confining, with a consequence of having transformed the participating energy quanta. The consequence manifests in the first law of thermodynamics addressing energy transformation. As far as the completed empirical observation is concerned, the first law has been formulated upon the elementary activity of measuring others in the first person status with its necessary consequence referred to also in third person description. Once we are determined to supplement the missing second person description, the on-going process of transforming the participating energy quanta can be made accessible descriptively in the present progressive tense.

6. EXPERIENCING, TRANSFORMING AND REPRESENTING

Second person description on the act of transformation now presupposes the activity of the first person status that enables itself to experience others. Energy transformation referred to in the first law of thermodynamics is upon experiencing temperature gradients more than anything else, since thermal equilibrium dismissing the likelihood of temperature gradients allows no transformation in the first place. The capacity of experiencing others addressable primarily in first person description is nothing other than about being sentient to influences coming from the surroundings.

In fact, any material bodies in the empirical world are involved in detecting and experiencing each other through mutual interaction. That is internal measurement (Matsuno, 1985, 1989; Rössler, 1987). Focusing on internal measurement in general or on the sentient capacity of molecules in particular may suggest a likelihood of describing what the underlying interaction may look like. Then we need to examine the possibility of an internal description of internal measurement. At this point, just for the sake of argument, let us consider an extremely simple case such that two bodies *A* and *B* are interacting with each other while reminding ourselves that there is no a priori guarantee for simultaneous coordination between *A*'s action upon *B* and *B*'s action upon *A* (Matsuno, 2002a). Internal measurement on the part of *A* implies that *A*, experiencing what has been presented by *B*, subsequently transforms the experience within itself such that as a consequence of the prior experience it will re-present its transformed self to *B*. *B* then experiences a new representation of *A*.

Similar processes are also occurring on the part of *B*. Since there is no third party guaranteeing an a priori coordination between the two bodies, the mutual interaction referred to as internal measurement proceeds indefinitely, repeating the cycle of experiencing, transforming and representing. Extension of the present scheme of internal measurement into the cases including more than two bodies would be straightforward as one applies the similar argument to an arbitrary binary pair chosen out of the whole array of interacting bodies.

One thus comes to notice that there is some room for representation even in the scheme of internal measurement. Although what is experiencing differs from what has experienced and similarly what is transforming differs from what has transformed, nonetheless, what is representing is identical to what has been represented, at least locally in time as referring to the available local representation because of its inertness. Of course, this local representation is temporary because it will constantly be updated by repeating the cycle of experiencing, transforming and representing on the part of any member of the participating material bodies (Hoffmeyer, 1996; Taborsky, 1997).

The significance of the occurrence of local representation, however, cannot be overemphasized. As referring to any local representation, one can envision a likelihood of such internal description in view of the fact that any description has to be anchored in something stationary or inert even temporaily. Otherwise, descriptive stability would be lost and no reliable description possible. Compared to external description of an invariable universal acting as a global coordinator, internal description grounds it only upon empirical local interactions, and their constant updating is inevitable. Continual updating of local representations in turn provides a means for describing a variable or changing object.

In fact, an impetus for initiating each cycle of experiencing, transforming and representing at every material body would be any inconsistency experienced among local representations presented from the neighboring bodies, with the tentative goal of precipitating local representations having no further inconsistencies (Matsuno, 2002a). At the same time, no material body can live with inconsistencies indefinitely as being left alone. Unless there is a prior coordination to eliminate those inconsistencies from the aggregate of local representations, each material body must constantly transform its experienced inconsistencies into an inconsistency-free local representation, despite the fact that it will subsequently come to meet further inconsistencies with yet others down the line. Material bodies surviving in the empirical domain would be only those that can even temporarily transform those experienced inconsistencies into inconsistency-free representations.

Local representations met with in internal descriptions remain stationary in between adjacent updates. One of the descriptive means to refer to such local representations would be some metric (Conrad, 1993), since metric would have no capacity to initiate changes on its own. Once a certain numerical figure is read out of an apparatus measuring whatever object, it will remain unchanged until the reading is further updated. An interesting example of this sort that we can encounter in the biological realm is the cell processing material resource flow (Pattee, 1982).

Material flow through the cell is a local representation of the cell. It is also subject to material flow continuity because no biological organisms can create material resources out of nothing. This condition of material flow continuity can give rise to a serious inconsistency among intervening local representations. Suppose that one cell happens to increase the intake of material resource flow so as to fulfill material flow continuity by diverting a portion of a similar flow to an adjacent cell. This diversion would then cause violation of material flow continuity on the part of the adjacent cell if left unattended. Since no violation of material flow continuity is allowed for local representations, the adjacent cell is forced to update its local representation so as to recover the condition of material flow continuity there (Matsuno, 1989). Inconsistencies among local representations are inevitable in the scheme of internal description, while no inconsistencies are allowed to survive in a globally consistent record approachable by external description. It is not a local representation but only the cell itself that can tame these inconsistencies to the tolerable level internally.

The biological cell as a material body comes to experience inconsistencies with the neighboring local representations in the form of the potential danger of violating material flow continuity. It then transforms itself from an inconsistency-experiencing body into an inconsistency-free body; otherwise it would fail to survive. The inconsistency-free body manifests itself as a representation being in accord with the condition of material flow continuity. What is unique to the cycle of experiencing, transforming and representing is that neither experiencing nor transforming can be represented as such. Although one cannot represent what these two operations look like in descriptive terms, they are certainly operative. The cell may experience the possible danger of violating material flow continuity, which external description cannot properly address as a matter of principle. The cell is intensive in activating itself to transform the experienced inconsistencies into the inconsistency-free representation.

Both experiencing inconsistencies among the available descriptive representations and exerting certain intensities for eliminating these inconsistencies from within are unique to internal description. On the other hand, once external description of an invariable universal is adopted, as has been common in the practice of empirical sciences, especially in physics, there could be no such things as experiencing inconsistencies residing within the established description, or exerting intensity from within. Everything would have to be globally consistent from the start by way of descriptive stipulation. This stipulation of external description is incontestable as an established methodology and remains perfectly legitimate. At the same time, internal description allowing for both inconsistent local representations and responsive intensities to ameliorate them would also be legitimate insofar as its descriptive characteristics are faithfully observed. The relationship between external and internal description is not antagonistic. Internal description is more encompassing. External description of an invariable universal can be precipitated from internal description if the latter is further constrained by a stipulation demanding no inconsistent representations and no fluctuating intensities from within. In other words, inconsistent local representations and intensities to ameliorate the inconsistencies are prior in our empirical world (Matsuno, 2002a; Gunji, 1995). Compared to external description of an invariable universal that is necessarily complementary to dynamics, internal descriptions being capable of precipitating external description of variable universals furnishes our languages with the capacity of enclosing dynamics.

In essence, what is required for describing dynamics is the descriptive capacity of crossing different grammatical tenses because internal description accommodates into itself both the local present perfect and the present progressive tenses. Any molecule is experiencing others nearby in the form of impinging intensities because it is sentient to the neighborhood events registered in the local present perfect tense. It is this sentient capacity of molecules which connects the local present perfect to the present progressive tense. Intensities experienced by a molecule derive from a constellation of events in the local present perfect, which does not necessarily guarantee internal consistency on the spot. If an intensity to be experienced by any molecule were globally consistent in the sense that the global aggregation of events in the local present perfect tense reduces to those registered in the global present perfect tense in a consistent manner, the principle of mechanics would apply. This implies that every movement of any molecule registered in the global present perfect tense could be quantifiable extensively in its spatio-temporal dimensions without directly referring to the intensive capacity on the part of molecules. Otherwise, a molecular capacity of being sentient to various intensities would have to be faced directly. Inconsistencies registered in the local present perfect tense induce intensities to be experienced by a molecule as a sentient being.

The molecular capacity of being sentient to others cannot properly be addressed in the standard third person description in the present tense because of the first person status of that capacity. Likewise, the capacity of transforming itself as responding to experiencing others cannot be referred to in third person description because of the lack of indexical activity in the latter, but can be done in second person description. In contrast, any molecule as a quantum can be equated to representation of a local character, though whose update would become inevitable. What becomes significant in this regard is that a quantum can serve as a representation of material origin in the repeated cycles of experiencing, transforming and representing. Any quantum in the making is accessible only locally in second person description in the present progressive tense. This stipulation now invite us how to construct a bridge connecting that quantum in the making to something globally consistent addressable in third person description in the present tense.

7. REPRESENTATIONS OF MATERIAL ORIGIN OR THE GENETIC CODES

The contrast between the global and the local will become most evident in quantum dynamics even in material terms alone (Matsuno and Salthe, 1995). Measurement in quantum dynamics is to project the state of one quantum system as an object onto the states of another quantum system serving as a measurement apparatus. That is a projection of one Hilbert space onto another Hilbert space as a form of internal measurement. This sort of projections is ubiquitous in the local Hilbert spaces supervening on the phase space, though the choice of the preferred basis set for each Hilbert space still remains unsettled (Everett, 1957; DeWitt, 1970).

The difference between one Hilbert space and the multiple supervening Hilbert spaces resides in the difference of the nature of interactions. When one focuses on a quantum in a given Hilbert space, the interactions responsible for forming the quantum are concurrent and synchronous among the constituent elements and do not require participation of interactions of sequential nature in time or measurement in short. For instance, a proton conceived as a quantum is a confinement of three quarks interacting concurrently and does not regard the quark interactions as being sequential in time. This distinction is however relative to the reference available and by no means absolute, although quantum phenomenon empirically reveals that there certainly occurs a robust confinement of concurrent interactions in the form of a quantum. The stipulation on the concurrent confinement of the quark interactions may be acceptable only to the extent that the resulting proton can remain robust. In contrast, the interaction of the proton with a nearby neutron is sequential in the referential framework that keeps the proton robust without suffering a nucleosynthesis with the neutron. Such an interaction of a proton with a neutron can be taken as an instance of internal measurement of projecting the influence from the neutron onto the proton sequentially in time. Projection of one Hilbert space onto another Hilbert space thus turns out inevitable once the indexical activity of referring to a quantum formed nearby is invoked as a material process of sequential nature.

The contrast between one local Hilbert space and the multiple local Hilbert spaces supervening on the phase space is ubiquitous in physics in general and in biology in particular. Covalent bonds making protein and nucleotide molecules can be taken as the quantum mechanical objects belonging to one Hilbert space, while non-covalent bonds making both protein folding and RNA folding operate between adjacent Hilbert spaces. A quantum as a concurrent confinement of internal measurement inevitably accompanies and allows for non-vanishing leakages of internal measurement out of the then available synchronous confinement, for instance, as letting the formation of non-covalent bonds be due to asynchronous leakages out of the synchronous confinement of interactions in the form of covalent bonds. Projection of one Hilbert space onto another one as a sequential process is just about the inevitable asynchronous leakage of internal measurement out of the quantum as its synchronous confinement. Sequential interactions summarized in the form of internal measurement will now have their unique implications and properties.

When the individual projections running over the supervening Hilbert spaces are considered to be mutually independent of each other because of the open-ended accumulation of splitting and bifurcation of the projections, the result could be seen as a collection of decohered quantum states belonging to one covering Hilbert space. The source of decoherence might be sought in the interaction with the environments (Gell-Mann and Hartle, 1990), though the present scheme may further beg the question of how such environments could be conceived in the first place.

One may approach the decohered quantum states statistically, with the consequence of quantum statistical mechanics reducible from quantum mechanics. However, this would not be the only possibility. One more possibility arising from the cumulative projections of one Hilbert space onto another is to establish a closed loop of projections in the sense that internal measurement letting one quantum system project itself onto another quantum system can form a closed loop among the participating quantum systems or local Hilbert spaces. Once such a closed loop of projections is formed, the quantum systems involved in the loop could remain coherent between them. Moreover, if the loop remains robust against the possible decoherences between the constituent members, such a loop could physically materialize in the phase space.

Suppose that there is a set of local Hilbert spaces H_i $(\mathrm{i} = 1, 2 \ldots n)$ for quantum systems S_i supervening on the phase space, and that system S_{i+1} internally measures S_i as applying the projection operator P_i onto an eigenstate or eigenvector of unit length ψ_i belonging to S_i. Then, a closure of the projections could appear if the projection P_n, unique to the combined quantum system S_n and S_1, when applied to ψ_n, happens to project the state in the local Hilbert space H_n back into the space H_1. The occurrence of such a closed loop of projections can conveniently be expressed as

$$P_n P_{n-1} P_{n-2} \ldots P_1 \psi_1 = \alpha \psi_1,$$

where α is a complex number representing the amplitude of the composite projections. This is a consequence of internal measurement between the local Hilbert spaces supervening on the phase space around a closed loop. As far as internal measurement is concerned, there can be no inconvenience even if the complex amplitude of the vector in a local Hilbert space is directly referred to. Since internal measurement constantly assumes the similar measurements to indefinitely follow internally, it would certainly admit participation of entangled quantum states as a consequence of each internal measurement. However, this form of quantum entanglement does not survive in external measurement.

External measurement of a closure of the projections can now give it a unique identification if the closure remains robust against any disturbances coming from the external apparatus for measurement placed outside of the whole supervening Hilbert spaces. If the closure remains robust against external measurement, it can also survive repeated external identifications. Once external measurement of a robust closure of the projections repeats itself frequently many times in an unbounded manner, the resulting complex amplitude of the eigenstate involved in the closure would become cumulative and reduce to the amplitude of the composite projections α raised to the power of the repetition number of external measurement. When the repetition number is ν, the complex amplitude of the composite projections due to external measurement repeated ν times would reduce to α^{ν} in terms of the amplitude of the composite projections α due to internal measurement. The consequential result is extremely constraining in that only the closure maximizing the absolute value of the amplitude of composite projections will effectively survive and all of the other alternatives will be wiped out if the repetition number ν is allowed to increase without limit. This is because the amplitude squared is proportional to the probability of occurrence in external measurement. In fact, the repetition number ν can be made arbitrarily large since the way of doing external measurement is completely separated from internal measurement.

Predominance of the closure maximizing the amplitude of composite projections may further proceed if the basis set characterizing each local Hilbert space is allowed to vary with the aid of interactions with the neighboring similar spaces. Since external measurement tends to focus on the largest amplitude of composite projections through its repetition, those variations on the basis sets that can increase the amplitude altogether could be selected for and fixed in the closure. When viewed from the perspective of external measurement, the basis set of each participating local Hilbert space may have an internally adaptive scheme of fixing its preferred basis through the closure of internal measurement toward increasing the amplitude of composite projections. An important point to note here is that the transtemporal identity of the agencies involved in identifying the closure is assumed by internal measurement, instead of external measurement. If external measurement assumed the transtemporal identity of identification, one would have to implement something additional to supplement material dynamics in general or quantum dynamics in particular for the sake of securing the observing agency in a transtemporal manner.

One necessary condition for the closure of projections to survive is that the similar closure of projections in the reverse direction has the smaller amplitude compared with that in the forward direction as letting it be suppressed infinitesimally small in external measurement in comparison to the former. The present absence of reversibility in projections may certainly be the case if temperature dynamics in the presence of temperature gradients is put in place. When there are two alternatives of the closure of projections, the clockwise and anti-clockwise circulation, the closure to be fixed will be the one that can mitigate the temperature gradients faster (Matsuno and Swenson, 1999). In any case, all biological phenomena proceed in the presence of temperature gradients in one form or another; otherwise the fate of ending up with heat death in thermal equilibrium would have to be unavoidable. In particular, the closure facilitating the mitigation of the available temperature gradients faster turns out to have the larger amplitude of composite projections between the two alternatives. The closure of composite projections to be met in biology thus breaks the reversibility in projections because of the de facto participation of metabolic energy flow of degradation in a unidirectional manner.

Fixation of the actual closure of composite projections common in the biological realm does not presume the agential capacity on the part of external measurement as frequently quoted in the form of the collapse of the quantum mechanical wavefunction imputed to the external observer. The contrast between the global and the local can thus become prominent once biology gets started (Matsuno and Salthe, 1995). External measurement on the global scale, though inevitable in the practice of empirical sciences in third person description, turns out simply an a posteriori consequence of internal measurement on the local scale addressable only in second person description. There is no need for scrutinizing the nature of the external observer, if any. The contrast between the global and the local is prevalent, but the characters of the descriptive tense to be referred to are different between the two.

At issue will be how to make a bridge connecting the present progressive to the present perfect tense, whichever materially or linguistically. In this regard, classical mechanics is exceptional in letting the present perfect tense be the precursor agency presiding over how the present progressive tense subsequently would proceed. Classical mechanics lets the present perfect tense run the present progressive tense by a decree. Quantum dynamics, on the other hand, grounds itself upon the present progressive tense being concurrent with the present perfect tense. This stipulation is entirely empirical. What is more, chemistry as a material substrate of biology is both empirical and linguistic in bridging the present progressive and the present perfect tense. Chemical reaction is about the process of updating the reactants once registered in the present perfect tense again through their resurrection in the movement in progress. Chemistry practices quantum dynamics so that the material record registered in the present perfect tense may constantly be updated. The impetus to the update is linguistically addressable at the least through the descriptive activity of fulfilling the principle of the excluded middle in the transference of the multiple agencies in the present progressive to the single agency in the past progressive tense. There is no chance of fulfilling the principle in the presence of multiple descriptive agencies in action. Only when the process is transferred into the past progressive tense, the principle would come to be observed because of the stipulation upon the single authorship controlling the whole descriptive enterprise practiced there.

Chemical reactants involved in the closure of projections on the supervening Hilbert spaces can now have their unique significance in their temporal or evolutionary development. The reactants residing in the innermost closure of projections can assume a special role in the movement in progress. The reactants involved in the outer closures are subject to the influences coming from those participating in the innermost closure since the innermost closure is part of the outer closures, while the reactants appearing only in the outer closures may not necessarily be the inevitable component for forming the innermost one. One thus notes that the chemical reactants residing in the innermost closure of projections could serve as the genetic materials such as RNAs and DNAs as the relics from the past that can direct or at least influence the chemical reactions currently in progress.

8. QUANTUM DYNAMICS PRECIPITATING THE GENETIC CODES

Quantum dynamics upholding the infrastructure of the material world is quite singular in distinguishing the present perfect tense from the present progressive tense. Material significance of precipitating the present perfect tense in the form of a quantum is in an almost complete confinement of internal measurement reverberating coherently in its inside. The fact that the confinement is almost complete instead of being literally complete rests upon the interplay between the covering phase space and the Hilbert spaces supervening on the former. Complete confinement of internal measurement in a quantum is adequately represented in a Hilbert space, while the representation that comes to experience similar representations of material origin available from the neighborhood Hilbert spaces can transform itself accordingly. Once the robust closure of the cycle of experiencing, transforming and representing is precipitated in the form of a quantum in the integrated sense, that quantum can serve as a historical and evolutionary record of material origin registered in the present perfect tense.

In particular, the origin of the genetic codes can be associated with the emergence of a robust closure of experiencing, transforming and representing in material terms. The act of representing is a projection bringing some attribute of a material object into the foreground as leaving the remaining behind in the background. The activity of such projection is unquestionably of material origin. How could such a separation between the foreground and the background be done is totally up to the material bodies involved. The present material aspect of the act of representing has no direct relevance to what the external observer or the physicist is used to do, since the observer cannot make any access to whatever material body without referring to its representation contrived for the purpose in the first place. This representation of exogenous origin has nothing to do with the representation that an arbitrary material body comes to internally induce upon other material bodies in the neighborhood. Nonetheless, the genetic codes are exceptional in that both the external observer and the chemical reactants involved in the robust closure of projections can happen to share the common representation. The genetic codes to the external observer are an irreducible representation of biochemical reactions proceeding in biological organisms, while they are reducible to the activities of experiencing, transforming and representing on the part of the participating chemical reactants from within as the internal observers.

Once the contrast between the external and the internal observers is focused, it will become evident that any dynamics conceived and formulated by the external observer cannot go beyond the stipulation that whatever framework of external representation, once contrived and accepted, remains immutable since then. This is the inevitable limitation that the external observer would have to admit. From the perspective of the externalist representation, the emergence of the genetic codes may look as a spontaneous appearance of de novo scheme of representation, which would have no common denominator with the preceding representations available to nonliving physical world. There could be no bridge connecting the two different schemes of representation in a coherently continuous manner. The emergence of the genetic codes may be taken as a spontaneous event going beyond reasonable comprehension to the mind of the external observer. In contrast, the chemical reactants as the internal observers are constantly involved in updating the internal representation available to them. The emergence of the genetic codes to the participating internal observers must be an event continuous to the preceding stages with even no biological organisms in sight since the internal representation has innate capacity of changing its own implication as going through the cycle of experiencing, transforming and representing repeatedly.

What makes the issue of the emergence of the genetic codes most intriguing has been our long-held tradition on sticking to a preferred choice of the fundamental predicates to represent the material world. Needless to say, the external observer, whether the philosopher or the physicist, or whoever else for this matter, has a privilege to choose a preferred set of the fundamental predicates. If the preferred scheme of representation has any empirical significance whatsoever, it must remain robust also in the empirical domain. In fact, our material world is equipped with the capacity of making itself robust enough. That is the robust cycle of experiencing, transforming and representing on the part of the participating material bodies from within. The external representation of empirical significance certainly remains robust in the empirical domain; otherwise there would be no chance of representing it as such empirically. The internal representation as a matter of fact comes to uphold the external representation, and not the other way around.

Once quantum dynamics is practiced and appreciated in second person description in the present progressive tense, the internal representation upholding the external representation may come to naturally be focused upon. The emergence of the genetic codes may also come to terms with material dynamics ubiquitous in the empirical world.

REFERENCES

Conrad M (1993) The fluctuon model of force, life and computation: a constructive analysis. Appl Math Comp 56:203–259

DeWitt BSM (1970) Quantum mechanics and reality. Phys Today 23(9):30–35

Everett H (1957) Relative state formulation of quantum mechanics. Rev Mod Phys 29:454–462

Gell-Mann M, Hartle JB (1990) Quantum mechanics in the light of quantum cosmology. In: Zurek WH (ed) Complexity, Entropy and the Physics of Information, Addison-Wesley, Redwood City, CA, pp 425–458

Gunji Y-P (1995) Global logic resulting from disequilibration process. BioSystems 35: 33–62

Hoffmeyer J (1996) Signs of Meaning in the Universe. Indiana University Press, Bloomington Indiana

Küppers B-O (1995) Die Kontextabhängigkeit biologischer Information (English transl. The context-dependence of biological information). In: Kruger L, Falkenburg B (eds) Physik, Philosophie und die Einheit der Wissenschaften. Grundlagen der exakten Naturwissenschaften, Band 10. Spektrum, Akademischer Verlag, Heidelberg pp 260–277

Matsuno K (1985) How can quantum mechanics of material evolution be possible? Symmetry and symmetry-breaking in protobiological evolution. BioSystems 17:179–192

Matsuno K (1989) Protobiology: physical basis of biology. CRC Press, Boca Raton Florida

Matsuno K (2000) The internalist stance: a linguistic practice enclosing dynamics. Ann New York Acad Sci 901:332–349

Matsuno K (2001) Thermodynamics as an interface agency tailoring quantum mechanics for the beginning of biology. In: Diebner HH, Druckrey T, Weibel P (eds) Sciences of the interface, Genista Verlag, Tübingen Germany, pp 143–154

Matsuno K (2002a) Evolving life: constant turnover from inconsistencies to intensities. In: Tymieniecka AT (ed.) Analecta Husserliana LXXIV, Kluwer, Dordrecht, the Netherlands, pp 215–231

Matsuno K (2002b) Beyond representation: bridging the chasm between the different grammatical tenses. In: Meyer M, Paton R (eds) Theoria et historia scientiarum 7 Special issue on knowledge, representation and interpretation – from quanta to cultures, Nicolas Copernicus University Press, Torun, pp 105–119

Matsuno K (2003) Quantum mechanics in first, second and third person descriptions. BioSystems 68:107–118

Matsuno K, Salthe SN (1995) Global idealism/local materialism. Biol Philos 10:309–337

Matsuno K, Swenson R (1999) Thermodynamics in the present progressive mode and its role in the context of the origin of life. BioSystems 51:53–61

Pattee HH (1982). Cell psychology: an evolutionary approach to the symbol-matter problem. Cognit Brain Theor 5:325–341

Rosen SM (2004) Dimension of apeiron: a topological phenomenology of space, time and individuation. Rodopi, Amsterdam

Rössler OE (1987) Endophysics. In: Casti JL, Karlqvist A (eds) Real brains, artificial minds, North Holland, New York pp 25–46

Taborsky E (1997) The textual society. University of Toronto Press, Toronto

CHAPTER 4

GENETICS, LIFE AND DEATH

Genetics as providing a definition of life and death

MICHEL MORANGE

Centre Cavaillès, Ens, Paris, France

morange@biologie.ens.fr

Abstract: There were two different and partially successive attempts of geneticists to associate genes with a definition of life. The first has its origin in the theoretical considerations elaborated at the end of the 19th century by biologists such as Hugo de Vries and August Weismann, looking for the molecular bases of biological processes and the mechanisms of their reproduction. It reached its most elaborate form in Hermann Muller's contributions, and pervaded genetics during the first part of the century. The second was the paradoxical result of the program of gene reification endorsed by molecular biologists, which progressively ruined the previous ambitions. This gave way to a less naive vision of the relation between genes and the definition of life, focused no longer on the materialistic description of the gene, but on its power to control the adaptation of organisms to their environment.

It is interesting that the link between life and death that has remained a constant of philosophical investigations from Aristotle to Bichat has kept its place in this genetic approach to life. Definitions of life by both the molecular and population genetics have tried – with less success for the former – to justify the place of death in the economy of nature. Models, derived from genetics, are now used to describe the transmission of behaviors and beliefs between humans. They give genetic models a strong visibility in contemporary thinking

1. INTRODUCTION

Genetics is one of the most pre-eminent sciences at the beginning of the 21st century. Other contributions in this book will explore the impact it has had on contemporary thinking and practice. My own contribution aims to show that this impact is also the consequence of the fact that genetics is, among the biological disciplines, the most tightly associated with a definition of life.

My project is rather ambitious, and two preliminary caveats concerning the meaning of "genetics" and "definition" are required. The ambiguous place of

A. Fagot-Largeault et al. (eds.), The Influence of Genetics on Contemporary Thinking, 51–60.

genetics among biological sciences, and in particular its recent fading with the emergence of new disciplines – post-genomics, proteomics – have been already discussed by philosophers and historians, without receiving simple explanations (Gayon, 2004; Morange, 2004). It is obvious that, despite its influence, genetics does not fulfill the criteria demanded of a discipline. Nevertheless, we will overlook this difficulty to define genetics: we do not intend to be more demanding than the editors of this book!

The second difficulty concerns the possibility of a definition of life. The philosophical debate on this question is traditional, and quite active today (Machery, 2005). We will adopt the minimal requirement for a definition: to establish a limited list of properties which are necessarily associated with any living object, and sufficient to distinguish it from inanimate matter. Within the framework defined by this limited ambition, we will show that the properties of the genetic material are first on the list for many biologists, the most narrowly associated with a possible definition of life. Our objective is not to estimate whether biologists are right or wrong to consider that the genetic properties of organisms are their most fundamental characteristic, and whether some of them – not all – have the right to consider that they have access through these properties to a definition of life, but to analyze when such a link between genes and life was elaborated, and how it evolved.

As we will see in the conclusion, the properties of the genes that have been selected for such a definition of life can easily be extended to behavior and mental activity. This explains the impact that models derived from genetics have indirectly today in fields quite remote from it.

2. THE GENES AS THE ATOMS OF BIOLOGY

It has been well demonstrated by historians that the roots of genetics are to be found more in the attempts of theoretical biologists to conceptualize a theory of heredity in the last decades of the 19th century than in the breeding experiments of Mendel (Olby, 1966, 1979). The focus of Mendel's work on the rules governing the production of stable hybrids – a traditional problem for breeders and crop producers –, and the weak impact his work had on contemporaries, even if it was not totally ignored by them, explain this limited influence.

Beyond the search for the mechanisms responsible for the reproduction of the characteristics of organisms from generation to generation, what was at stake at the end of the 19th century in this theoretical work was the construction of a materialistic and corpuscular basis of living processes (Pichot, 1999; Allen, 2000); it aimed at determining what was specific to organisms, and what distinguished them from inanimate objects.

The expression "atoms of biology" to designate the genes became highly popular among geneticists very soon after the "re-discovery" of Mendel's laws, and the rapid development of a genetic science. Genes were initially considered as a convention of writing, in the same way as chemists used atoms during the first part of the 19th century, at a time when they did not believe in their material existence.

This comparison between the language of genetics and the language of chemistry did not help to dissipate the ambiguity of the expression "atoms of biology", because atoms were later shown to be real objects. The comparison was made more explicit by the Harvard geneticist William Castle: "All observed inheritance phenomena can be expressed satisfactorily in terms of genes, which are supposed to be to heredity what atoms are to chemistry, the ultimate, indivisible units, which constitute gametes much as atoms in combination constitute compounds" (Castle, 1919; quoted by Allen, 2003). But ambiguities still remained: are genes the atoms of biology simply because they represent one of the fundamental constituents of living organisms, one of the constituents that have a major role in controlling the properties of the organism? There is probably more in this expression than mere recognition of the importance of genes: genes are the atoms of biology because they are to the living world what atoms are to the inanimate one; the foundation from which its properties are derived. The comparison made by Muller between the mutations induced by the experimenter – by chemicals or X-rays – and the transmutation of matter – lead into gold – so desperately sought by alchemists (Muller, 1927) is not only a fashionable metaphor, exploited to outline the powers reached by the new biology. It also demonstrates how seriously the expression "atoms of biology" was considered in the case of genes.

The fundamental place of genes in the living world was strengthened by the historical prospect outlined by Muller (Muller, 1929): since genes are so closely associated with life, the origin of life was probably closely linked with the invention of genes. This hypothesis supported one of the two lines of research which dominated the origin-of-life field of research after its theoretical emergence in the 1920s – with the simultaneous and independent studies of Oparin and Haldane –, and its brilliant experimental burst at the end of the 1940s – with the *in vitro* synthesis by Miller of amino acids, the building blocks of proteins, in conditions mimicking those of the primitive Earth (Kamminga, 1988). The search for self-replicating RNA molecules is the present avatar of this line of research.

Another independent but not antagonistic way to tighten the link between genes and life was to invest genes with functions considered as fundamental for life. Troland took the first and major step in this direction by identifying genes with enzymes (Troland, 1917). Since the most fundamental characteristic of organisms was their capacity to catalyze the chemical conversion of molecules in conditions in which the chemist was helpless, genes themselves had to be catalysts, enzymes, the central catalysts within the cells giving rise to the other cellular enzymes.

The enzymatic conception of life dominated genetics between 1930 and 1950, before giving way to the informational conception of life with the characterization of the DNA molecule, the demonstration of its coding function, and the rise of molecular biology (Olby, 1974). But the replacement of the enzymatic conception of the gene by an informational one did not abolish the central functional role the gene had in the cellular economy. The reverse was true: most cellular activities were conceived as resulting from exchanges of information, for which the gene was the source and the foundation.

It is relatively easy to understand why, in this context, the characterization of gene – DNA – structure was considered as the discovery of the secret of life. One understands also why the question "What is life"? disappeared in the 1960s and 1970s. This disappearance was obviously a "legacy of molecular biology" (Shostak, 1998): the question was no longer asked since a satisfactory answer had been obtained. Life is the consequence of the presence of a genetic information, and of a genetic code allowing this information to be translated. Some philosophers were not far from accepting the answer proposed by molecular biologists: the presence, at the basis of life, of a "logos" explained why language was so apt to open the way to an understanding of life (Canguilhem, 1968).

Why did this vision not resist the critiques, and progressively gave way to a more subtle relation between genes and the definition of life? A first answer emerges simply from the historical sketch that we have drawn. The rapid and abrupt shift from an enzymatic conception of life to an informational one was not satisfactory: it could not abolish this privileged link between life and metabolism, the well-proven existence of active chemical processes and exchanges within organisms, and between organisms and the environment. The informational conception of life could not fully replace the enzymatic one. In addition, the progressive reification of the gene, its transformation into a precise molecular structure throughout the 20th century, somehow opposed the central role given to the gene. The functional identification of the gene with the organism (Ravin, 1977) remained possible as long as the structure of the gene was unknown. When its structure was described, it was no longer possible to attribute to the DNA molecule the dynamic and "living" properties which were attributed to the abstract form of the gene. A new way to link the existence of genes to a definition of life had to be elaborated.

3. GENES ARE ESSENTIAL FOR THE ACTION OF NATURAL SELECTION

Among the different contemporary definitions of life which are given by biologists, one of the most frequent is that what characterizes organisms is their capacity to "reproduce imperfectly". This definition includes two components, both essential: the power to reproduce, characteristic of life; and the fact that this reproduction is sufficiently faithful to allow a correct functioning of the progeny, but not perfect, not preventing the appearance of slightly different forms of organisms, likely to be better adapted to their environment. Imperfect reproduction is the condition required for the action of natural selection, and the historical development of life.

An outstanding question is to know whether this definition is sufficient. Many other objects in the universe, in particular artefacts created by man, are able to reproduce imperfectly: some computer programs have this capacity. Are they alive, or will they become alive in the near future? It is a matter of lively debate whether any material support can accommodate life (Adami, 1998).

One frequently refers to this kind of definition as a Darwinian definition of life. Is this definition totally independent of the existence of any genetic mechanism?

The answer is obviously no. The action of natural selection is only possible if a minimal system of heredity exists, allowing the variations appearing at each generation to be – at least partially – transmitted. Richard Lewontin described in its simplest form this minimal genetic requirement for life: (1) Different individuals in a population have different morphologies, physiologies and behaviors ("phenotypic variation"). (2) Different phenotypes have differential rates of survival and reproduction in different environments (differential adaptation). (3) There is a correlation between parents and offspring for their contribution to future generations (adaptation is heritable). It is interesting that Lewontin does not precisely describe the mechanisms of heredity; all that counts is that at least part of the differential adaptation is transmitted (Lewontin, 1970).

Because genetic mechanisms are the main mechanisms of heredity to have been progressively forged and retained by natural selection, they are fundamental in this Darwinian definition of life. As sophisticated as they are – with the very precise copy at each generation of the long protein sequences –, they perfectly provide the basis on which natural selection can efficiently play its role.

However, what matters is the existence of a mechanism of replication of variations, not the precise way it works. This disconnection between what is needed for an efficient action of natural selection, and the nature of the mechanisms involved, has two important consequences (Rosenberg, 1994). The first is that any characteristic, which has been shown to be reproduced more or less faithfully, can be supposed to be the target of natural selection; even if it is difficult or even impossible to provide any information on the way these characteristics could be "genetically transmitted". This explains why population genetics has progressively extended its scope far beyond the structural characteristics of organisms. Life histories – the complex and diverse trade-off between growth, maintenance, reproduction and ageing – became the preferred subject of study of ecologists and geneticists in the 1960s (Korfiatis and Stamou, 1994). In the 1970s, behavior was the focus of attention of behavioral scientists and geneticists with the development of sociobiology. It was not considered necessary to have unraveled the complex ways by which genes might control processes as complex as ageing and altruistic behavior to study the strategies used by organisms to increase their fitness by modifying these processes.

Lack of interest in the mechanisms underlying the hereditary transmission of these characteristics is paradoxically also a strength, because it liberates these studies from an attachment to a specific mechanism of heredity. If tomorrow epigenetic mechanisms of heredity, linked to reversible modifications of chromatin and DNA, were shown to have a dominant role in the transmission of such or such hereditary characteristic, the models would have to be slightly modified, but the ambition to explain the adaptation of organisms to their environment would remain unchanged.

Genetic mechanisms – whatever they are – are the handles that natural selection can grip. Without these handles, natural selection would have no strength. Genetics is therefore an integral part of the definition of the most fundamental characteristic of organisms – their power to adapt and to evolve.

4. THE SEARCH FOR A MECHANISTIC LINK BETWEEN GENES AND DEATH

The relations of genes to death evolved in parallel to this search for the place of genes in the definition of life, although the situation remains less clear even today. The close and antagonistic relations between death and life, and the necessity to understand the role of the former in telling us something significant about the latter, are a full part of the philosophical tradition.

Investigations into whether the duration of life is genetically-controlled – and the search for the genes involved – were initiated as soon as the laws of genetics were re-discovered. Lucien Cuénot was the first to isolate lethal forms of genes in mice (Burian and Gayon, 1999). Experimental genetic studies of the duration of life – on *Drosophila* – developed as early as the mid-1920s (Pearl et al., 1923; Gonzalez, 2003). At the end of the 1960s, when the notion of genetic program became fashionable, the existence of a genetic program of aging and death was immediately proposed, and widely accepted. This program was an extension of the genetic program of development, its last instructions. This notion of a genetic program of death was initially presented in a positive way by François Jacob in *The Logic of Life. A History of Heredity* (Jacob, 1970), a book which contributed to the popularity of the notion of genetic program. Jacob abandoned this notion of a genetic program of aging in his next book *The Possible and the Actual* (Jacob, 1981), probably because it had been heavily criticized by evolutionary biologists.

These criticisms did not prevent some molecular biologists from looking for the genes controlling the duration of life. The use of genetic engineering technologies and animal models allowed in the 1990s what had not been possible in the 1920s: the isolation of the genes involved, and the characterization of their functions (Guarente, 2003). These studies are still rapidly developing, and so far it is difficult to appreciate the significance of the results. Two important characteristics of these genes were utterly unforeseen: the number of these genes seems limited, and their role in the control of the duration of life has been conserved during evolution. To delay the process of senescence in these animals now seems to be a proximate goal for biologists. The manipulation of these genes in the human species raises wonderful or frightening prospects, depending upon the point of view.

Despite its apparent success, such a vision of genes controlling the duration of life – and death – was rapidly challenged both by cell biologists and population geneticists. The former showed the existence of a specific form of cell death – by apoptosis –, totally different from the form of cell death which results from the absence of nutrients and oxygen occurring when the organisms die. During apoptosis, cells play an active role in their own death – whence its name of programmed cell death.

The first observations of apoptosis did not question the existence of a genetic program of death or its link with a developmental program, since they were made on the massive cell death process that accompanies metamorphosis, a specific phase of development for insects and amphibians (Lockshin and Williams, 1964, 1965). But as findings accumulated on this specific form of death, it appeared less and less linked with the program of development and more and more related to the capacities of organisms to adapt to changing environments. It is the absence of cell death, not its occurrence, which in most cases constitutes a threat to the organism.

The existence of two different deaths – programmed cell death and death of the organism – and their apparent independence remains a puzzling observation for biologists (Kwang-il Kang and Morange, 2001). Recent reports have tried to establish links (Cohen et al., 2004; Kujoth et al., 2005), but these remain tenuous.

5. THE ROLE ACCORDED TO GENES IN AGING AND DEATH BY EVOLUTIONARY THEORY

The strongest challenge to the idea that genes control aging – within a genetic program or not – came from evolutionists. August Weismann had been the first to look for an evolutionary explanation of aging and death at the end of the 19th century – see Klarsfeld and Revah (2000) for a history. This path was followed by Peter Medawar (Medawar, 1952) and George C. Williams (Williams, 1957). The present model proposed by population geneticists is that the duration of life (and the occurrence of death) is not actively controlled by genes, but that specific forms of genes have been progressively introduced by chance during evolution and can affect the duration of life, and increase the probability of death simply because they do not sufficiently affect fitness to be eliminated by the action of natural selection. Two slightly different possibilities remain: these mutations affect the organisms too late, at a time when, in natural conditions, most have already succumbed by accident; or their negative effect is balanced by the positive effect they have on reproduction (antagonistic pleiotropism). A later modification has been made to the evolutionary theory of aging and to the "antagonistic pleiotropism" model by appealing to the "disposable soma" theory (Kirkwood and Rose, 1991): the organism differentially allocates a fixed number of resources to reproduction and maintenance.

In this evolutionary vision, genes are also active players in aging and death. But in contrast to the role they have in the adaptation of organisms to their environment – where they are the handles on which natural selection can act –, their role in death derives from a partial escape from the action of natural selection. Life and death constitute the two opposite faces of the complex relations between natural selection and the genetic apparatus.

This evolutionary theory of aging – which preceded the invention of a genetic program of death – has since obviously swept the former away. As we have seen, it has not prevented the huge efforts directed at the isolation of genes whose variations are involved in the tempo of aging. The possible relations between programmed cell death and the death of organisms, and the conservation of aging genes during evolution, are observations that somehow challenge the evolutionary theory of aging. The present situation is unstable: nevertheless, whatever way the balance shifts, the direct or indirect link between genes and death will persist.

6. CONCLUSION: GENES AND MIND

Genes are also linked with the highest expressions of life, behavior and mind. Once again, genes and behavior have been linked in a variety of ways over the last hundred years. Traditionally, genes and the manifestations of mind were linked by considering that behavior and mental activity were somehow "under the control" of genes. What this expression meant precisely was unclear, and very diverse observations, research and models came under this fuzzy headline: the conviction that most (70%) human genes participated in the formation of the brain, or the possibility that some behavior is directly controlled by one gene, could be easily accommodated in this vision. Human genome sequencing – and the discovery of the low number of human genes – was clearly a blow to the idea that human higher activities are directly reflected in the complexity of the genome, whereas the isolation of master genes controlling behavior has frequently been disappointing, and has lost its attraction – possibly with the exception of the *fruitless* gene of *Drosophila* (Manoli et al., 2005). Gene variations can have a dramatic effect on behavior, but this does not mean that the genes that are involved are behavioral genes. However, the possibility that the human mind consists of cognitive modules that evolved through the pressure of natural selection still leaves a place for the genes in the construction of the mind (Buller, 2005).

But the relation between genes and mind can be seen from a very different point of view. Ideas could be transmitted and inherited in the human species in a way similar to that of genes. Richard Dawkins was the first to give this model a precise form with the invention of "memes", the equivalent of genes for mental activities (Dawkins, 1976). The proposal to compare the evolution of ideas to the evolution of organisms, and to give a similar role to natural selection in both processes, is not new: it was clearly expressed by Jacques Monod in the last chapters of *Chance and Necessity* (Monod, 1970). Although the comparison made by Richard Dawkins between memes and genes has not met with full success, many research programs presently aim to apply the complex models elaborated by evolutionary geneticists to the transmission between humans of knowledge, behaviors and beliefs. As before, these models are Darwinian. They do not refer to the genes of geneticists, but somehow apply the same corpuscular model behind the success of genetics to the transmission of mental states. From mind to life and death, genes occupy a dominant position.

REFERENCES

Adami C (1998) Introduction to artificial life. Springer-Verlag, New York

Allen GE (2000) The reception of Mendelism in the United States, 1900–1930. C. R. Acad. Sci. Paris, Sciences de la vie 323:1081–1088

Allen GE (2003) Mendel and modern genetics: the legacy for today. Endeavour 27:63–68

Buller DJ (2005) Adapting minds: evolutionary psychology and the persistent quest for human nature. MIT Press, Cambridge, MA

Burian RM, Gayon J (1999) The French school of genetics: from physiological and population genetics to regulatory molecular genetics. Annu Rev Genet 33:313–349

Canguilhem G (1968) La nouvelle connaissance de la vie: le concept et la vie *in* Etudes d'histoire et de philosophie des sciences, Vrin, Paris pp 335–364

Castle WE (1919) Piebald rats and the theory of genes. Proc Natl Acad Sci USA 5:126–130

Cohen HY, Miller C, Bitterman KJ et al. (2004) Calorie restriction promotes mammalian cell survival by inducing the SIRT1 deacetylase. Science 305:390–392

Dawkins R (1976) The selfish gene. Oxford University Press, New York

Gayon J (2004) La génétique est-elle encore une discipline? Med/Sci 20:248–253

Gonzalez BM (1923) Experimental studies on the duration of life: VIII. The influence upon duration of life of certain mutant genes of *Drosophila melanogaster*. Am Natur 62:289–325

Guarente L (2003) Ageless quest: one scientist's search for genes that prolong youth, Cold Spring Harbor Laboratory Press, Cold Spring Harbor, NY

Jacob F (1970) La logique du vivant, Paris, Gallimard. Trans (1974) The logic of life, Princeton Scientific Library, Princeton, NJ

Jacob F (1981) Le jeu des possibles, Paris, Fayard. Trans (1982) The possible and the actual, University of Washington Press, Seattle, WA

Kamminga H (1988) Historical perspective: the problem of the origin of life in the context of developments in biology. Orig Life Evol Biosph 18:1–11

Kang K, Morange M (2001) Succès et limites de l'étude moléculaire de la mort cellulaire programmée. Annal. d'Hist. du Vivant 4:159–175

Kirkwood TBL, Rose MR (1991) Evolution of senescence: late survival sacrificed for reproduction. Phil Trans Roy Soc Lond *B* 332:15–24

Klarsfeld A, Revah F (2000) Biologie de la mort. Odile Jacob, Paris

Korfiatis KJ, Stamou GP (1994) Emergence of new fields in ecology: the case of life history studies. Hist Phil Life Sci 16:97–116

Kujoth GC, Hiona A, Pugh TD et al. (2005) Mitochondrial DNA mutations, oxidative stress, and apoptosis in mammalian aging. Science 309:481–484

Lewontin RC (1970) The units of selection. Annu Rev Ecol System 1:1–18

Lockshin RA, Williams CM (1964) Programmed cell death – II. Endocrine potentiation of the breakdown of the intersegmental muscles of silkmoths. J Insect Physiol 10:643–649

Lockshin RA, Williams CM (1965) Programmed cell death – I. Cytology of degeneration in the intersegmental muscles of the *Pernyi* silkmoth. J Insect Physiol 11:123–133

Machery E (2005) Why I stopped worrying about defining life . . . and why you should as well, privileged communication

Manoli CM, Foss M, Villella A et al. (2005) Male-specific *fruitless* specifies the neural substrates of *Drosophila* courtship behavior. Nature 436:395–400

Medawar PB (1952) An unsolved problem of biology. H. K. Lewis, London

Monod J (1970) Le hasard et la nécessité, Paris, Le Seuil. Trans (1972) Chance and necessity, Vintage books, New York

Morange M (2004) La génétique: une science aux caractéristiques très particulières. Bull Hist Epistém Sci Vie 11:235–247

Muller HJ (1927) Artificial transmutation of the gene. Science 66:84–87

Muller HJ (1929) The gene as the basis of life. Proceedings of the 4th international congress of plant science, Ithaca, I:879–921

Olby RC (1966) Origins of Mendelism. Constable, London

Olby RC (1974) The path to the double helix. Macmillan, London, Chap 9

Olby RC (1979) Mendel no Mendelian. Hist Sci 8:53–72

Pearl R, Parker SL, Gonzalez BM (1923) Experimental studies on the duration of life: VII. The Mendelian inheritance of duration of life in crosses of wild type and quintuple stocks of *Drosophila melanogaster*. Amer Nat 57:153–192

Pichot A (1999) Histoire de la notion de gene. Flammarion, Paris

Ravin AW (1977) The gene as catalyst; the gene as organism. Stud Hist Biol 1:1–45

Rosenberg A (1994) Instrumental biology or the disunity of science. The University of Chicago Press, Chicago
Shostak S (1998) Death of life: the legacy of molecular biology. Macmillan, London
Troland LT (1917) Biological enigmas and the theory of enzyme action. Am Natur 51:321–350
Williams GC (1957) Pleiotropy, natural selection, and the evolution of senescence. Evolution 11: 398–411

PART II

GENETICS AND PHILOSOPHY OF SCIENCE: THE REDUCTIONISM DEBATE AND BEYOND

CHAPTER 5

MOVING BEYOND THE INFLUENCE OF MOLECULAR GENETICS ON THE DEBATE ABOUT REDUCTIONISM IN PHILOSOPHY OF BIOLOGY

FREDERIC BOUCHARD
Department of Philosophy, University of Montreal, Montreal, Canada
F.bouchard@umontreal.ca

Abstract: The rise of molecular genetics has had a fundamental influence on the development of philosophy of biology because of its central role in the debate concerning reductionism. I begin by describing notable attempts to reduce Mendelian genetics to molecular genetics, the lightning rod of reductionist attempts in philosophy of biology. I then suggest that the syntactic reductionist may wish to focus on neglected biological cases (e.g. ecosystem evolution) that *may* eventually yield laws that could provide easier reduction than the attempted reduction in genetics

1. INTRODUCTION

A belief common among philosophers and biologists alike is that Mendelian genetics has been or is in the process of being reduced to molecular genetics, in the sense of formal theory reduction current in the literature.

(Hull, 1972, 491)

The debate about reductionism in genetics was joined by a number of philosophers (...) Unlike so many debates in philosophy, this one has yielded a near consensus. The dominant view is that classical genetics has not, is not, and will not be reduced.

(Waters, 2000, 539)

What happened in the intervening thirty years between these two accurate depictions of the state of play in philosophy of biology shows the influence the study of molecular genetics has had on philosophy of biology: it has shaped in radically different ways its consensus views about reductionism. In this paper, I will trace some of the impact the rise of molecular genetics has had on philosophical debates, mostly on the issue of reductionism.

A. Fagot-Largeault et al. (eds.), *The Influence of Genetics on Contemporary Thinking*, 63–80.

The problem of reductionism has taken many forms in the philosophy of biology literature. As Waters (2000) points out, most debates concerning reductionism in philosophy of biology fall into variants of one of three debates. Are genes, whatever they are, the main or sole unit of natural selection? Is Mendelian genetics reducible to molecular genetics? If so, can molecular genetics lead to a reduction of biology to chemistry and physics –if not in practice at least in theory–? In this article I will focus on the second question, for it might be the best way to understand just how much the development of molecular genetics has influenced debates in philosophy of biology.[1]

I will also focus mainly on reductionist approaches that wish to use the second question as a way to bolster claims about the third question. In other words, the reduction of Mendelian genetics to molecular genetics is often (but not always[2]) used as the first step to a wide reduction of biology to chemistry and physics.

I will describe how the rise of molecular genetics first motivated syntactic reductionists and their attempt to reduce Mendelian genetics to molecular genetics (Schaffner, Ruse), and then motivated anti-reductionists to reject positivist accounts of biology altogether (Hull). I also briefly address non-positivist accounts of reduction that arose from this debate (Wimsatt, Sarkar). After this examination of some of the foundational aspects of contemporary philosophy of biology, I will offer a possible way out for syntactic reductionists; reduction and syntactic approaches to biology might be available outside of molecular genetics. The last section sketches this possibility with the case of potential ecosystem evolution (Van Valen) and the laws it may inspire. This programmatic point is intended as an exhortation for reductionists to move beyond molecular genetics and examine other types of biological explanations.

2. GENERAL STATE OF PLAY

Arguably, the received view in philosophy of science for the first half of the 20th century was to endorse some variety of positivism concerning the nature of scientific theories. Physics was seen by many as the paradigmatic example of a scientific discipline where an axiomatic structure girded the theoretical apparatus and correspondence rules linked the logical structure to the body of empirical observations. Although there have been notable arguments to show that physics itself was a poor exemplar for a positivist view (e.g. Cartwright, 1983), most philosophers of science saw physics has following the positivist programme somewhat faithfully.

Biology did not seem to fit the hypothetico-deductivism mold. Contra rare but notable attempts to axiomatise certain aspects biology (see Woodger et al., 1937,

---

[1] All the varieties of reductionism (and counter-moves) that have arisen in philosophy of biology cannot be adequately addressed in this format. For an excellent survey of all the positions see Wimsatt, 1979; Schaffner, 1993; Sarkar 1998. What I intend to do here is describe the reductionist impetus molecular genetics has provided to some philosophers of biology.

[2] Waters 2000 for instance argues that, epistemically, Mendelian genetics is reduced to molecular genetics but he does not believe that such a reduction is but the first step in a metaphysical reduction of biological phenomena to organic chemistry, chemistry and ultimately physics.

1952; Ruse, 1973; Williams, 1970; Rosenberg, 1985), most philosophers of science argued that biology was better understood as a special science whose object was too context dependent to lead to the uncovering of universal laws that could lead to the axiomatisation of biology.

Not surprisingly many biologists – who after all had vested interests in the issue – were vocal in their take on the rise of molecular genetics (e.g.Crick, 1966; Monod, 1970; Dobzhhansky, 1966; Mayr, 1961; Simpson, 1964; see Beatty 1990 for a useful analysis of many biologists' anti-reductionist response). But aside from the obvious repercussions the development of molecular genetics had on biological sciences, the molecular discoveries of the 1940s, 1950s and 1960s had lasting consequences in the humanities has well. In this article, the focus is on the influence the rise of molecular genetics had on philosophical reductionist and anti-reductionist stances, and not on biologists' take on the significance of this scientific discipline.

The influence of molecular genetics on philosophical discourse is striking not only for its scope but for its dialectical nature: first, with the development of molecular genetics, reductionist approaches for biology seemed intuitively vindicated in part because of the provisional pre-analytic reductionism defended by Crick (1966), Monod (1970) and other architects of the revolutkion in molecular genetics.[3]

The reductionist hope in philosophy of science was that Mendelian genetics could be reduced to molecular genetics: it was assumed pre-analytically that the Mendelian generalizations concerning the similarity between parents and their offspring could be explained by molecular mechanisms. The problem of reductionism in philosophy of science being usually articulated as a problem concerning relations between theories (not entities), the question is whether Mendelian genetics *theory* could be reduced to molecular genetics *theory*. Most philosophers of biology may well grant that biological 'stuff' is ultimately reducible to physical 'stuff'; metaphysically the commitment of the vast majority of philosophers of biology is to physicalism. The disagreement is over whether or not this commitment entails that our theories should all be reducible to the basic physicalist theories (i.e. physics).[4] The anti-reductionist claim is that biological theories are not reducible to physical theories.[5]

The development of molecular biology has proven to be the blessing of various reductionist projects. It has been a blessing for it seemed to warrant both a metaphysical reductionism (all biological processes could in theory be reduced

[3] I'm including Monod as a pre-analytic reductionist because of his public advocacy for the overwhelming explanatory force of molecular genetics, even though some of his remarks would clearly disappoint reductionist philosophers. A clearly reductionist manifesto 'Ainsi définie, la théorie du code génétique constitue la base fondamentale de la biologie' (Monod, 1970, 12) is immediately followed by the qualification that '*Ce qui ne signifie pas, bien entendu, que les structures et fonctions complexes des organismes puissent être déduites de la théorie, ni même qu'elles soient toujours analysables directement a l'échelle moléculaire.*' (ibid.).

[4] See Rosenberg, 1985, Chapter 4 for more on this issue.

[5] An example of this weak physicalist position can be found in Kitcher 1984.

to processes at the macro-molecular level and then to chemical and physical processes) and a methodological reductionism (science should focus on macro-molecules to make the largest strides in explanation of the biological world). The near omnipresence of DNA as the inheritance mechanism in nature, its high fidelity and its role in the presence of variation in populations seemed to give biology strong chemical and physical underpinnings making the reduction, replacement or elimination of biological theories in favor of chemical or physical theories only a matter of time (this enthusiasm is evident for example in Schaffner, 1969, Ruse, 1971a, 1971b, Rosenberg, 1985). Although reduction, replacement and elimination are not identical, proponents often share the goal of exchanging a higher-level of organization explanation for a more fundamental level of organization explanation.[6] When reduction fails, we can either assume that replacement or elimination is the way to go or assume that we have genuine emergence (or, we can, obviously, question both theories' adequacy). Pre-analytically, molecular genetics seemed to provide the foundation for reduction in biology.

This initial reductionist enthusiasm was quickly dashed: opponents of reductionism in philosophy biology quickly pointed out that the canonical model of reduction (Nagel, 1961) did not apply in the case of the purported reduction of Mendelian genetics to molecular genetics. The most lasting argument was provided by Hull (1976, more on his argument below), with an influential addition provided by Kitcher (1984). What seemed as a blessing turned out to be a curse for the seemingly promising reduction of Mendelian genetics to molecular genetics has in fact not occurred and various attempts have shown the seams of theories of reduction. The apparent failure of this reduction prompted many (if not most) philosophers of biology to reject the approach that fueled the desire for reduction in the first place namely post-positivism, *at least* about biology but probably about other scientific disciplines as well.

In some respects, reductionist and anti-reductionist arguments were both inspired by developments in molecular genetics and as such Watson and Crick's discovery and other aspects of the development of molecular biology shaped much of the discussions in philosophy of biology from the 60s on. The attempt to reduce Mendelian genetics to molecular genetics has therefore had the strange dual role of encouraging post-positivists reductionist projects for biology (e.g. Schaffner, 1969; Ruse, 1971a, 1971b; Waters, 1990; Rosenberg, 1985 or in a more intuitively vein Crick, 1966 and Monod, 1970) while providing the theoretical foundation to reject positivist projects altogether (Hull, 1976; Beatty, 1980; Kitcher, 1984; Lloyd, 1994; Thompson, 1989 and many others). Ironically molecular genetics is seen by both provincialists and autonomists as the best reason to adopt their respective position.[7]

---

[6] Wimsatt gives us a helpful general criterion for identifying a reductionist: 'A reductionist conceptual scheme (or world) is at least one in which when explanations are not forthcoming in terms of other same-level entities and phenomena, one is more likely to look for (or find) an explanation in terms of lower-level phenomena and entities than in terms of higher level phenomena and entities.' (Wimsatt, 1976, 689).

Last but not least, the philosophical study of molecular genetics has lead to new research programmes to better understand reduction outside of post-positivist strictures (Wimsatt, Sarkar). As such, the influence of molecular genetics is almost omnipresent in philosophy of biology. Let us now see in more details how the reductionism debate was shaped by molecular genetics.

3. THE SYNTACTIC HOPE

The standard bearer of syntactic reductionism in biology has been Schaffner who with two seminal papers (Schaffner, 1967, 1969, and an amended version in 1993) attempted to show that a variety of the Nagelian reduction model obtained (or *could* in theory obtain) with the reduction of Mendelian genetics to molecular genetics.

Reduction as described by positivists is a type of explanation: the theory being reduced is explained by the reducing theory by virtue of it being deduced from the reducing theory. In the classic Nagelian account theory T2 is reduced to theory T1 when all the generalizations made by T2 can be logically deduced from the generalizations made by T1. In cases where such deduction is not available because the theories don't share a common language in which the deduction could make 'sense', T1 can be supplemented by 'translating' principles (or bridge laws) that can make sense of claims made by T2.

In Schaffner a significant accommodation is introduced, for such 'straight-forward' deductions would be very rare. He recognizes (Schaffner, 1967, 1969, 1993) that Mendelian genetics (the T2 in this case) as is cannot be reduced to molecular genetics (T1 here), but he suggests that we still have a relatively 'Nagel-like' reduction (hereafter nomological reduction) if we allow for T2 to be 'slightly' modified prior to reduction. T2 will not be reducible to T1 without the intermediate introduction of new terms and concepts. In other words, although Mendelian genetics (T2) is not deducible from Molecular genetics (T1), a slightly modified version of Mendelian genetics (now T2*) that takes into account evidence marshaled by T1 will ultimately reduce T2, or T2* (see Schaffner, 1969, 332–333, 1976 and 1993, 429). Of course, not all possible modifications of T2 will be acceptable; if that were the case, reduction would be as trivial as adding ad-hoc modification to any theory to permit deducibility. T2 and T2* need to be 'strongly analogous' for the reduction to be acceptable. More will be said later about this concept of analogy.

The translation problem comes in part from the fact that a radically different concept of gene is involved in T1 and T2. Mendelian genes are identified via function whereas molecular genes are identified structurally.[8] Since function does

[7] Rosenberg (1985, Chapter 2) tags 'provincialists' as those who see biology as merely a province of physics (i.e. biological theories can ultimately be reduced to Physical theories), whereas 'autonomists' see biology consistent with but independent from theories in physics.

[8] Molecular genes as Waters (1994, 2000) points out are also defined functionally, but in a way that is constrained by the chemical–physical structure of the gene.

not have a necessary connection to specific structures, it isn't clear how Mendelian genes could be made to correspond to molecular genes. A Mendelian gene can only correspond to a disjunction of molecular genes and as such, no deduction will be straightforward. Reductionists (following Benzer, 1957) suggested that T2 be modified into T2* so as to replace the 'vague' term of Mendelian gene into one of three categories: mutons correspond to the smallest unit of mutation, recons are the smallest unit of recombination and cistrons are the smallest unit of phenotypic expressions. Cistron is in some case analogous to specific Mendelian genes. By replacing a functional term of Mendelian genetics by terms that can be translated into structural terms (e.g. the cistron represents the minimum number of nucleotides that codes for a given phenotype.), the reduction of now a somewhat structural theory (T2*) to structural theory (T1) seems more feasible. The problem remains of providing a criterion for distinguishing strong analogy and weak analogy: how are we to know whether there are too many differences between T2 and T2* to really claim that a Nagel-like reduction can be effected? The necessary introduction of the concept of analogy might be the hint that a syntactic nomological conception of reduction is not appropriate, minimally in the case of genetics, but probably in cases other than genetics as well.

Ruse offers a related but somewhat different analysis of the relationship between Mendelian genetics and molecular genetics (1971). The position slightly changed over the years, but the original argument was in favor of 'strong replacement' of Mendelian genetics by molecular genetics, as opposed to 'weak replacement' more in line with the reduction discussed above. This replacement is justified for various reasons: among them are the fact, Ruse argues, that molecular biology explains everything that Mendelian genetics explained, explains things that Mendelian genetics cannot explain, and explains things that contradict explanations made by Mendelian genetics. For our purposes the salient point is that the relation is one of replacement because it isn't *exactly* fitting with Nagel's model of reduction. Ruse argues in 1971a, 1971b that the conditions for Nagelian reduction are almost met but not exactly – the introduction of the concept of analogy makes this evident. Ruse argues for an *informal* reduction of Mendelian genetics to molecular genetics, one that strictly speaking is actually a relation of replacement but that is close enough to reduction to be considered such in everyday practice. Ruse later (1976) adopted an account closer to Schaffner's, explaining that the informal nature of the reduction is merely the result of the uncompleted status of the enterprise. In some sense, Schaffner and Ruse while not *yet* having provided the full nomological reduction, argue that, based on partial reductions, optimism concerning the eventual reduction of Mendelian genetics to molecular genetics is warranted. But this optimism needs to be justified, especially in light of the constraints that a realistic appraisal of the mechanisms involved provides.

The relationship between Mendelian genetics and molecular genetics is what has been called a many to many relation. A single Mendelian gene does not correspond to a single molecular gene (and the same is true for loci and alleles) Hull describes how this type of relationship dooms reductionist projects:

Phenomena characterized by a single Mendelian predicate term can be produced by several different types of molecular mechanisms. Hence, any possible reduction will be complex. Conversely the same type of molecular mechanism can produce phenomena that must be characterized by different Mendelian predicate terms. Hence reduction is impossible.

(Hull, 1976, 39)

Moreover, Hull points out that the lack of search of derivations in the actual practice of molecular biologists – coupled with the multiple realizability problem described above – puts into question the optimism displayed by the would-be reductionists.

Hull's pointing out of the many-many problem is not the only problem for the reductionist. As Hull also pointed out (1972, 1974, 1976) Schaffner and Ruse did not in fact provide a full description of the theories involved in the reduction. Even the axiomatic nature of existing 'laws' in biology needs profound qualification: Kitcher (1984) pointed out how it was recognized early on that Mendel's 'laws' of heredity were not as universal as one could hope and so there wasn't a unified coherent Mendelian theory of genetics.[9] A related problem with reduction is that it isn't even clear that there is a unified molecular gene concept that could be the reducing concept for the Mendelian gene concept. Many have concluded that 'gene' in molecular genetics "is a place holder or dummy term designating a motley collection of different kinds of molecular entities" (Waters, 2000, 544). This has motivated Waters to suggest a unified concept of molecular gene (see Waters, 1994 and a revised argument in 2000) that according to their account accurately describe most biologists' usage of the term. Waters argues that a coherent structural concept of gene is available.[10]

So the first step to obtain a nomological reduction is still missing in action: what are the theories involved in the reduction? Although attempts to provide axiomatic descriptions of certain aspects of biology have been made (e.g. populations genetics and evolutionary theory: Ruse, 1973; Williams, 1970; Rosenberg, 1985) they remain more promissory than complete and as such, the propedeutical step of actually identifying the full theories involved in reduction in biology has not been successfully completed.

This problem is compounded if we adopt Schaffner peripherality view concerning reductionism (Schaffner, 1974). He acknowledges that the reduction problem is

[9] Many have since argued that while Kitcher may be right about the *initial* laws he is wrong about the present 'anomalous' status of Mendelian theory. As Sarkar (1992) puts it: 'it is apparent that classical genetics does have a theory: the laws of classical genetics are simply Mendel's laws corrected by a host of discoveries, mainly in the 1920s, long before the advent of molecular biology.'(Sarkar, 1992, 185). Sarkar goes to argue that actually it is molecular genetics that might still lacking unified theory (not Mendelian genetics).

[10] Waters provides what is arguably the most convincing molecular gene concept that may enable eventual reduction: 'Despite the appearances, however, there is a clear and uniform way of understanding genes at the molecular level, which (...) can be summarized as follows: a gene g for linear sequence l in product p synthesized in cellular context c is a potentially replicating nucleotide sequence, n, usually contained in DNA, that determines the linear sequence l in product p at some stage of DNA expression.' (Waters, 2000, 544).

a problem concerning mainly philosophers not biologists: since initially offering his reduction model, he observed that deducibility or logical coherence is nowhere apparent in the strategy deployed by molecular geneticists. So, not only is nomological reduction of Mendelian genetics to molecular genetics 'irrelevant' to biologists as Schaffner concedes, but as Hull (1972, 1974, 1976, 1979) points out it cannot satisfy philosophers for it remains incomplete. Since both Schaffner and Hull agree that a reductionist approach is not an adequate descriptive approach concerning biology, the question then becomes whether to take a reductive normative approach seriously. But with Schaffner's account, not only reductionism is not normative for biologists, but because of its promissory nature it is not evident why it should be a normative account satisfactory for philosophers either.[11]

The Schaffner model would not satisfy biologists, cannot satisfy anti-positivists, and will leave post-positivists waiting for the *actual* reduction to be provided. This being said, Schaffner's reduction model remains a significant contribution to the debate. Not only it provided the first thorough attempt to reduce Mendelian genetics to molecular genetics but it did so in a way very close to Nagel's reduction model, hinting that a post-positivist approach to biological knowledge, if not easy, was more than a pipe dream. By introducing the role of analogy, Schaffner might have weakened the positivist credentials of his account, but he showed that limited accommodations might keep the syntactic account of theories relevant for biology. Analogy does not fit nicely in the syntactic view, but it provides a way to accommodate some discontinuity within scientific theorizing while not giving up on the enterprise of rational reconstruction and of nomological reduction.

Ruse's later endorsing of Schaffner's model has the advantage of minimizing the peripherality point. Ruse wishes to show how adopting such a view may in time provide real discoveries (Ruse, 1976). By doing so, Ruse may convince some to adopt the normative claim even tough the reduced and reducing theories have not been fully described. The difficulty of providing complete theories of both Mendelian genetics and molecular genetics as convinced some to give up on nomological reductionism. Even initially enthusiast reductionists (e.g. Rosenberg, 1985) later bit the many-many bullet: 'A more reasonable response is to accept that the systematic deduction of theory that characterizes physics and chemistry is not to be had in biology, to explain why not, and to show that the absence of reduction is no threat either to materialism or to the unity of science.'(Rosenberg, 1994, 22)[12] If, as Rosenberg (1994) concedes, selection

[11] Many philosophers of biology do not think that a reductive approach is a good normative approach for biology in general: 'Perhaps a completed science would be able to unite biology and physics but this claim about some hypothetical future says nothing about we should conduct our investigations in the present' Sober, 1993, 25. As Rosenberg points out (1985, 88), the reductionists often adopt a normative stance because their position is steeped in metaphysical commitments whereas most anti-reductionist's favoring of descriptive claims is grounded in epistemological presuppositions.

[12] The apparent failure of straight-forward reductionism has pushed Rosenberg to now embrace a larger role for ecology in evolutionary biology. See Bouchard and Rosenberg 2004 and Rosenberg and Bouchard 2005 for examples.

for function is blind to structure, maybe one should expect the supervenience of Mendelian genetics on Molecular genetics without the type of identity that would enable Nagel-like reduction. But Waters (2000) gives reasons to weaken the problem of multiple-realizability: he argues that in genetics we are not faced with the 'extreme' functionalism that is found in psychology. As Sarkar also points out (1992, 186), the acceptance of molecular genetics supervenience is premature. Basically they both show that the apparent immense disjunction of mechanisms involved in any genotype-phenotype relation is more limited than it is believed.

But Ruse (1976) had already offered hope that we weren't in fact faced with many-many problem in the first place:

> What seems to follow from the genetical case is not the conclusion that a deduction is in principle impossible, but that more information is required at the molecular level showing why the various phenotypic effects occur. Then we can get away from such correspondences as $\beta = (B1 \vee B2)$, replace them by such as $\beta 1 = B1$ and $\beta 2 = B2$, and hence deduction is once again possible
>
> (Ruse, 1976, 635–636)

The move is to show how we are not in fact faced with one molecule related to multiple phenotypes but rather different molecules involved with different phenotypes. The disjunction of phenotype is replaced by various identities making nomological reduction possible.

Such a complete molecular story was famously provided for the case of sickle-cell anemia (see Rosenberg, 1985, Chapter 6 for what is, in his own words, a 'triumph of reductionism'). But such complete descriptions are relatively rare (even today). Aside from the practical difficulty of providing such fine-grain molecular story to permit reduction to obtain, Hull, pointing out the same possibility – albeit with irony (1976, 20, 42) – warns us that such a move makes any reductive claim uninformative: we wish to relate kinds, not particulars. If the only way to get rid of the many-many problem is to over-specify the individual molecular mechanism in order to show that it can only lead to one phenotype, we will lose much of the explanatory appeal that laws had in the first place. Ruse (1976, 638) argues that Hull's point while valid may not be as extreme as Hull believes the kinds might be less ample in scope but we will not be relating solely particulars.

The worry remains that post-positivist approaches to reduction in the case of molecular genetics will be either impossible or at least horribly complex. Or that fundamental changes to the syntactic view may have to be made (e.g. introduction of analogy) to make reduction possible.

These difficulties motivated others with reductionist inclinations to offer a novel account of reduction not dependant on theories (or for that matter, not fully dependant on a syntactic account of scientific explanation). This departure from Nagel's account of reduction is in some sense the second phase of the attempt to reduce Mendelian genetics to molecular genetics and one example of it will now be briefly examined.

4. EXPLANATORY REDUCTIONISM

What the normative–descriptive question as well as the peripherality issue highlight is the tension between various forms of reductionism at play in this debate: Sarkar (1998) offers a thorough classification of types of reductionism. He offered a more basic classification in 1992 that will adequately serve our purposes here. Sarkar distinguishes between theory reductionism, explanatory reductionism and constitutive reductionism. Schaffner and Ruse's attempts to reduce Mendelian genetics to molecular genetics are clearly of the first kind (although because reduction is a form of explanation, it also belongs to the second kind[13]). Following post-positivist strictures, Schaffner and Ruse aim to deduce the laws of heredity provided by Mendelian genetics from the purported laws of molecular genetics (derived from laws of chemistry and physics). This 'gamble' fails in part because of the difficulty of identifying coherent theories and universal laws both in the reducing and reduced 'theories'.

But there remains a certain intuitive reductionism at play in the relationship between Mendelian genetics and molecular genetics, and Sarkar argues that some varieties of explanatory reductionism can account for it. Following a related argument provided earlier by Wimsatt (1976), Sarkar argues that there is a reduction involved here but a reduction of explanations (not theories). Wimsatt stressed that the Nagelian model (and its refinements with Schaffner) do not reflect how scientific progress is often achieved via the successful explanation of systems by their lower-level parts. In explanatory reductionism, reduced entities are explained by the reducing one. Most important to that approach is that those entities are not necessarily theories and laws but could also include empirical generalizations and even as Sarkar points out (1992, 170) individual observation reports. Since, as anti-reductionists of all kind gleefully point out, molecular genetics does not seem to have the unifying reducing theory that enable nomological reduction in the first place, reductionists might be better served giving up on theory reduction in molecular genetics and instead focus on explanatory reductionism advocated by Wimsatt and developed by Sarkar. What Schaffner's model made apparent is that modification of the theories involved may be necessary to effectuate the reduction. Wimsatt argues that this modification only shows how much theories co-evolve (Wimsatt, 1976, 682). If the theories are not atemporal unchangeable constructs, the reductionist may be better served focusing on more 'stable' entities such as explanations where piecemeal reductions can be obtained.

---

[13] Most reductionist projects fall into more than one camp: constitutive reductionisms are interested in making metaphysical claims, which is obviously what Schaffner is after. For our purposes the interesting comparison is between reductionist projects where only reduction between theories is allowed, and reductionist project where other theorizing 'tools' can be used as well. The latter is interesting for it introduces a type of reductionism that while sometimes compatible with positivism (when the reduction entities are theories) often does not have to be compatible (when the reduction entities are observations or imperfect generalizations).

The general framework of explanatory reductionism is to begin with an explanation of a phenomenon and identify all the causal interactions at a given level of organizations. Then we take the same explained phenomena and attempt to describe how it could be the result of interactions at a lower more fundamental level of organization. Maybe more importantly we analyze the causal chains in light of the higher level description (i.e. we recognize that the higher level explanation might still be useful to evaluate the claims made the lower level characterization). In the case of genetics this means that although we might reduce a trait's 'origin' to molecular factors, we may maintain the trait's identification thanks to the functional analysis provided at the level of classical genetics.

One obvious advantage of this approach is that it deflates some of Hull's (1974, 1976) original reproach concerning reduction: by widening the potential membership of reduction relation to include non-universal or non-perfectly axiomatised theories, explanatory reductionism can now better reflect actual biological practice, where some intuitive form of reductionism is practiced (mostly in molecular biology) outside of an explicit positivist framework. It also weakens Schaffner's peripherality claim concerning reductionism. Some explanatory reductionism might lead to new empirical claims therefore granting reductive explanation much instrumental value to scientists as well as philosophers.

But adopting this account of reduction may constitute a Pyrrhic victory. By adopting explanatory reductionism, the reduction of Mendelian genetics to molecular genetics is replaced by the project of reducing *some* explanations of Mendelian genetics to *some* explanations of molecular genetics, thereby jettisoning the reduction *en masse* developed in post-positivist projects.

5. NOMOLOGICAL REDUCTIONISM BEYOND MOLECULAR GENETICS

Where is a reductionist to go? If genetics does not seem to lend itself easily to axiomatisation, or to law-like description, or rather if one of the best candidates for nomological reduction faces seemingly insurmountable problems, is the lesson to be drawn, as it claimed by anti-positivists, that biological knowledge does not lend itself to a positivist account of scientific explanations? Maybe we should simply adopt, as Wimsatt, Sarkar, Waters and others suggest, a more pragmatic view of reduction more in line with biologists' actual practice, an explanation-based reduction. Although Schaffner, Ruse and others' approach faces many problems, it maybe too early to toll the bell for a post-positivist approach to biology.

What I would now like to suggest is that the nomological reductionist may wish to shift the focus of reduction away from molecular genetics onto other biological processes that might lend themselves to easier reduction to chemical or physical processes. Although some successes in piecemeal reduction to molecular genetics can be achieved if one endorses explanatory reductionism, Nagel-like reduction may still be available in other areas of biology. Admittedly, like other nomological reduction arguments examined here, this argument is more promissory than conclusive; it will be sketched merely as an invitation for further inquiry.

In the same way that many anti-reductionists decry the over emphasis on molecular genetics by reductionists, the reductionist should examine how the anti-reductionist consensus holds in face of attempts to axiomatise other fields of biology. Few areas of biology have taken the question of reductionism as seriously as ecology.

As Hagen (1989) explains, two diverging traditions have driven its development in the last century. This distinction stems from the ecologist Hutchinson (1978) who distinguished two theoretical camps in ecology: the *merological* approach is more demographic and examines populations of independent organisms (Simberloff is a notable 'merologist'). *Holological* approaches 'focus upon the flow of materials and energy through ecosystems without considering the organisms that are constituents of the system' (Hagen, 1989, 434). The idea of ecosystems as super-organisms stems from that approach. On the population ecology side of ecology (Simberloff and others) a very sharp Ockam's razor is used to dismiss most apparent higher-level community interactions that holological minded ecologists (e.g. E. Odum) find interesting. Ironically, holological thinking may lead us to nomological reduction: the holological tradition has inspired some biologists to posit general trends if not laws about evolutionary processes.[14] I wish now to briefly examine Leigh Van Valen's interest in ecology and energy, only to suggest that the post-positivist may want to examine the possibility that axiomatisation and reduction may lie beyond molecular genetics.

If as Van Valen (1973) famously suggested, evolution is the control of development by ecology, the possibility that energy transfers could explain much of ecosystem interaction would have interesting consequences beyond ecology. In some respect this is exactly what Van Valen hypothesized in subsequent papers (Van Valen, 1989, 1991). Van Valen ultimately advocates an energetic paradigm of evolution, according to which *all* evolution is ultimately an attempt to increase control over available energy. This is compatible with the holological approach which as a 'physiological' approach is sympathetic to the reduction of many ecosystem interactions to chemical and physical processes, albeit at a very high level of organization. Ecology has been interested in energy transfers since its birth (Hagen, 1989) and it has also been interested in how this could explain evolutionary phenomena (e.g. Tansley, 1935). The choice of Van Valen is justified in part because he has thought about the nature and role of laws in scientific explanations (Van Valen, 1972) and he has had, in large part because of the Red Queen, a lasting impact in evolutionary theory work. For Van Valen, the data demanding explanation are the apparent constant extinction rates for given groups.[15] Van Valen suggests the Red Queen's Hypothesis to explain these patterns: in ecological terms,

---

[14] See McShea, 1998 and Bouchard, 2004 for some analysis of the usage of energy as a trend in evolution.

the environment for a given group is stochastically deteriorating at a constant rate. One should understand deterioration of the environment in adaptive terms: the environment at t2 (i.e. later time) has degraded from the environment you were adapted to at t1 (i.e. original time). If the environment is constantly degrading, one has to adapt simply to stay at the same 'coordinate' in a fluctuating adaptive landscape (defined in resource space), or as Van Valen quotes Lewis Carroll's *Through the looking Glass*: 'Now here you see, it takes all the running you can do, to keep in the same place' (Van Valen, 1973, 25, n.32). This hypothesis is intended to explain the apparent linearity of extinction curves across taxa.[16]

The problem is that 'perfect' linearity should not be expected: concave survivorship curves should be expected. Taxa occupying larger spatial area – in real space *not* in resource space – are harder to stamp out and therefore, older taxa are 'probably' harder stamp to out since they have had more time to increase the area they occupy. In other words, the linearity observed in the survivorship graphs, does not match ecological assumptions concerning the difficulty of younger, less spread out, taxa to persist: in fact the whole project makes the strong Markhovian assumption that present survivorship probability is independent from previous probabilities. As Van Valen points out, his data implies that 'The probability of extinction of a taxon is then effectively independent of its age' (Van Valen, 1973, 17).

After examining (and ultimately rejecting) various explanations to explain the linearity as merely data artifacts, Van Valen concludes that 'extinction in any adaptive zone occurs at a stochastically constant rate' (Van Valen, 1973, 16). This is actually a reformulation of what he sees as the new 'law' namely that 'The effective environment of the members of any homogeneous group of organisms deteriorates at a stochastically constant rate' (ibid). Since he does not believe his law is exceptionless, it is doubtful that his law is as robust as laws found in physics, but his principle can be seen as an attempt to provide a principle more general than any previous attempt. As such it is somewhat surprising that Van Valen's Red Queen Hypothesis, well known by philosophers of biology, has not been seen as the foundations for a syntactic argument.

In two subsequent articles, this resource-control space morphs into an energy-control space which is the salient point for our purpose here. The first paper 'Three paradigms of evolution' (Van Valen, 1989) describes the superiority of an energetic paradigm over a reproductive paradigm (the current consensus) for understanding evolution. In 'Biotal evolution: a manifesto' (1991) Van Valen describes a possible application of this energetic paradigm to a 'new' level of selection: Van Valen suggests that we could theoretically observe evolution of complete biotas. Although

---

[15] 'The method is an application of the survivorship curve of the population ecology (including demography). It is a simple plot of the proportion of the original sample that survives for various intervals. (...) A logarithmic ordinate, standard in ecology, gives the property that the slope of the curve at any age is proportional to the probability of extinction at that age.' (Van Valen, 1973, 1).

[16] As Van Valen notes however, even though extinction rates are constant over the majority of temporal scales, they are not constant over geological time (see Van Valen, 1973, 10–12).

Van Valen argues that one should look at biotas (all the living entities in a given area), not at whole ecosystems (which comprise the biota and all the non-living substances in an area), other comments he has made concerning the potential evolution of non-living systems (see Van Valen, 1989) warrants an interpretation of the 'biotal evolution manifesto' as an 'ecosystem evolution manifesto'). On could see biotas as super-organisms but the moniker as Van Valen points out himself is not appropriate. He is more comfortable with the concept of a community, but a community of different members (a motley crew different species, non-organic material etc.). He takes this biotal evolution view to argue for a new understanding of evolutionary theory in terms of energy control. These biotas, he argues, evolve but do not really reproduce. How can we make sense of their evolutionary success *sans* reproductive success? Van Valen argues that any system will try to control more free energy than its competitor. This attempt to control more energy might translate in having offspring (or it might not). But in the end, the currency of evolutionary success will *always be* energy commandeering. A sketch of this would be to say that ecosystem A is fitter than B if A controls more energy (Van Valen does want to use joules or some other unit here). Survival of the fattest if you will.

Part of Van Valen's motivation for proposing an energetic paradigm is his dissatisfaction with the lack of interaction between molecular genetics and ecology (Van Valen, 1989, 1). Van Valen argues that energy control is the only thing being maximized in nature by all species and that sometimes it translates into higher reproductive numbers, whereas sometimes it translates into higher growth (e.g. in asexual clonal species).

Although Van Valen has not joined this hypothesis to this Red Queen argument, one can see the possibility of finding a law of evolution, based on ecological considerations (holological considerations) and that could breath some life into post-positivistic accounts in biology.

6. POST-MOLECULAR REDUCTIONISM: EXAMINING ECOLOGY AND EVOLUTION

Taking seriously such bridges between evolutionary explanations and holological approaches may solve some of the outstanding issues reducing the plausibility of reductionism in philosophy biology. Looking back at our discussion concerning the first round of the reductionism fight in philosophy of biology, we identified a few issues that now have some hope of being solved.

Firstly, an energetic paradigm weakens the autonomist view of biology in favor a kind of provincialism. This is not a novel result of course: the holological approach in ecology has from the start been an effort to bring physiology, physics and chemistry in ecological research (see Hutchinson, 1978). But the novelty with Van Valen's extension of this approach is that we now have a vindication of provincialism that flows from ecology to evolutionary biology. If, as is often argued, evolution is the strongest source of unification in biology, the reductionist should look seriously at any attempts to reduce the assumed autonomy of biology: if some ecological processes can potentially be reduced to energy transfers, physical and chemical reactions, if some evolutionary explanations concerning ecosystems and their evolution can possibly be described in terms of energy transfers as well, and finally if evolution is to be the unifying framework for biology, one might wish to reconsider the autonomy of biological explanations.

Secondly, the criticism that reduction cannot obtain because the theories involved are fully fleshed out is not eliminated but weakened. As we have noted, a problem bedeviling the reduction of Mendelian genetics was that T1 and T2, the 'theories' in Nagelian reduction, have not been fully specified. This is still the case for an energetic paradigm of evolution but notice that Van Valen was explicitly searching for a new law of evolution and that the energetic paradigm is likely to be applicable in a very general fashion. This is not to say that the theory is completely specified but rather that there is hope that it could be. Using Van Valen's ideas concerning the Red Queen and biotal manifesto, one could describe a general theory evolution that could then be eventually reduced to a theory about energy (i.e. a physical explanation). This is very promissory, but I hope I have given enough reasons to justify further inquiry.

Third, Ruse's case for strong replacement may apply in this case as well; an energetic paradigm might be able to explain everything that current evolutionary accounts explain while explaining phenomena that cannot be accounted for by these same models (e.g. ecosystem evolution), thereby warranting a strong replacement of accepted explanations and theories.

Fourth, even if we do not flesh out T1 and T2 in the case of a reduction of biotal manifesto, we would probably obtain explanatory reductionism as it is described by Sarkar. Such reduction is, as he point out, already obtained in the case of molecular genetics, but it strengthens the appeal of a shift of focus to ecological cases: in the eventuality that we do not establish laws in an energetic paradigm of evolution, one may still provide many explanatory reductions focused on specific energy interactions in ecosystems and how they affect or are affected by evolutionary processes. Evolutionary processes that seemed non-reducible might be explained solely in energetic terms, bringing biological processes under the mantle of chemical and physical explanations.

Finally, the peripherality fear (Schaffner, 1974) is unwarranted in the case of ecology and evolution. Van Valen and other ecologists' approach allows for the explanation of previously unexplained or unrecognized phenomena such as biotal evolution. This type explanation will require partial reduction to physical and chemical explanations and is a true contribution to biology, and not just a philosophers' quarrel.

7. CONCLUSION

Much more would need to be said to be convinced of the appeal of these accounts to explain evolutionary processes, but the fact remains that these suggestions among others may provide the foundations for what could constitute biological laws. My suggestion here is nothing but programmatic: from the energetic camp in ecology and its possible application in evolutionary theory, we may see law-like statements that could lend themselves to eventual reduction in the spirit of Nagel or Schaffner's accommodation. Ironically, this suggestion was indirectly hinted very early in the molecular genetics reductionism debate. Ruse's 1971a, 1971b article 'Reduction, Replacement, and Molecular Biology' is an oft-quoted article concerning reductionism, but in the same *Dialectica* issue, Ruse pens another reductionist manifesto ('Two Biological Revolutions') that also deserves our attention. In it, Ruse points to some of the dividends molecular genetics has had on other areas in biology.

> These advances in molecular genetics have spurred and illuminated other branches of biology, outstanding amongst which is perhaps that which is normally called bioenergetics, the part of biology which concerns itself with the flow of energy through living systems.
>
> (Ruse, 1971a, 1971b, 22)

Here Ruse is interested in energy flows at the level of mitochondria and chloroplasts but the same reason that make bio-energetics appealing to a reductionist may make the large scale ecological study of energy transfers at the ecosystem level appealing as well: energy is a physical quantity transferred via chemical reactions in biological processes.

The immediate problem with such a suggestion is that, apart from the difficulty of identifying and measuring energy transfers for all biotas, the level of abstractness of the claims offered by such an account of biotal evolution only make the hypothetical reduction to physical theories even less useful to every-day biology, possibly condemning the reductionist achievement to the peripherality described by Schaffner. Although this worry is justified, it should not cloud the philosophical issue. If some aspect of biological knowledge does lend itself to axiomatisation, if such axiomatisation is in term compatible with theories in physics (in this case energy), then reductions closer to the Nagelian concept might be attainable. If that is the case, the positivists and their intellectual heirs might have a toehold in biology.

It is not that surprising that most philosophers of biology have not looked to holological approaches to bolster nomological reductionism: between the failure of the apparent best candidate for reduction (i.e. Mendelian genetics to molecular genetics), the almost complete control evolutionary biology has had over debates in philosophy of biology since the 1970s, the ever growing popularity of the merological approach in ecology, ecosystem evolution is not exactly on the philosopher's (or the biologist's) radar. But the recent reappraisal of group selection (Sober and Wilson, 1998), artificial selection experiments for whole ecosystems (Swenson et al., 2000a, 2000b) a reappraisal of the holological approach by philosophers is urgently needed. If such reconsideration leads to entertaining Van Valen's ideas about evolution, we may get laws in biology where it was assumed there where none.

Crick, Monod, and many others inspired some philosophers to adopt some form of reductionist outlook on biological processes. Although this molecular outlook has yielded positive outcomes, it hasn't vindicated the positivist hope for nomological reduction. I suggest here that reductionists might wish to look at other areas of biology as inspiration as well: I am not arguing that reductionists get over molecular genetics but rather that they stop ignoring the vast expanses of biological knowledge left out from that debate.

Molecular genetics inspired many philosophers to think about the very small to understand biological processes, what I am suggesting here is that it may be time to think very big in order to bring biology closer to chemistry and physics.

REFERENCES

Beatty J (1990) Evolutionary anti-reductionism: historical reflections. Biol. Philos 5:199–210

Beatty J (1980) What's wrong with the received view of evolutionary theory. Proceedings of the Biennial Meeting of the Philosophy of Science Association 397–426

Benzer S (1957) The elementary basis of heredity. In: McElroy WD, Glass B (eds) The Chemical Basis of Heredity. Johns Hopkins University Press, Baltimore, pp 70–93

Bouchard F (2004) Evolution, Fitness and the Struggle for Persistence. PhD dissertation, Duke University

Bouchard F, Rosenberg A (2004) Fitness, probability and the principles of natural election. British Journal of Philos. Sci 55 (4):693–712

Cartwright N (1983) How the laws of Physics lie. Oxford University Press, Oxford

Crick F (1966) Of molecules and men. The John Danz Lectures. University of Washington Press, Seattle

Dobzhansky T (1966) Are naturalists old-fashioned? Am Nat 100:541–550

Hagen JB (1989) Research perspectives and the anomalous status of modern ecology. Biology and Philosophy 4:433–455

Hull DL (1976) Informal aspects of theory reduction. Proceedings of the Biennial Meeting of the Philosophy of Science Association, 1974, 653–670

Hull DL (1974) Philosophy of biological science, Prentice-Hall Foundations of Philosophy Series, Prentice-Hall, Englewood Cliffs, NJ

Hull DL (1979) Reduction in genetics. Philos Sci 46(2):316–320

Hull DL (1972) Reduction in Genetics-Biology or Philosophy? Philos Sci 39(2):491–499

Hutchinson GE (1978) An introduction to population ecology. Yale University Press, New Haven, CT

Kitcher P (1984) 1953 and all that. A tale of two sciences. Philos Rev 93(3):335–373

Lloyd EA (1994) The structure and confirmation of evolutionary theory. Princeton University Press, Princeton backs, PrincPapereton, NJ

Mayr E (1961) Cause and effect in biology. Science 134:1501–1506

Mayr E (1982) The growth of biological thought diversity, evolution, and inheritance. Belknap Press, Cambridge, MA

McShea DW (1998) Possible largest-scale trends in organismal evolution: eigh live hypotheses. Annu Rev Ecol Syst 29, 293–318

Monod J (1970) Le Hasard Et La Nécessité Essai Sur la Philosophie Naturelle de la Biologie Moderne. Éditions du Seuil, Paris

Nagel E (1961) The structure of science problems in the logic of scientific explanation. Harcourt, Brace & World, New York

Rosenberg A (1994) Instrumental biology, or, the disunity of science. Science and its conceptual foundations. University of Chicago Press, Chicago

Rosenberg A (1985) The structure of biological science. Cambridge University Press, Cambridge, New York

Rosenberg A, Frédéric B (2005) Matthen and Ariew's obituary to fitness: reports of its death have been greatly exaggerated. Biol Philos 20

Ruse M (1973) The philosophy of biology. Philosophy: philosophy (Hutchinson Firm). Hutchinson, London
Ruse M (1976) Reduction in genetics PSA: Proceedings of the Biennial Meeting of the Philosophy of Science Association, 1974, 633–651
Ruse M (1971a) Reduction, replacement, and molecular biology. Dialectica 25(1) 39–72
Ruse M (1971b) Two biological revolutions. Dialectica 25(1) 17–38
Sarkar S (1998) Genetics and reductionism, Cambridge Studies in Philosophy and Biology. Cambridge University Press, Cambridge, UK
Sarkar S (1992) Models of reduction and categories of reductionism. Synthese 91:167–94
Schaffner KF (1967) Approaches to reduction. Philos Sci 34:137–147
Schaffner KF (1993) Discovery and explanation in biology and medicine. The University of Chicago Press, Chicago
Schaffner KF (1974) The peripherality of reductionism in the development of molecular bioloy. J Hist Biol 7:111–129
Schaffner KF (1969) The Watson-Crick model and reductionism. Br J Philo Sci 20:325–348
Simpson GG (1964) This view of life: the world of an evolutionist, 1st edn. Harcourt, Brace & World, New York
Sober E (1993) Philosophy of biology. Dimensions of philosophy series. Westview Press, Boulder, CO
Sober E, Wilson DS (1998) Unto others the evolution and psychology of unselfish behavior, Harvard University Press, Cambridge, MA
Swenson W, Arendt J, Wilson DS (2000) Artificial selection of microbial ecosystems for 3-chloroaniline biodegradation. Environ Microbiol 2(5):564–571
Swenson W, Wilson DS, Elias R (2000) Artificial ecosystem selection. Proceedings of the National Academy of Science 97(16):9110–9114
Tansley AG (1935) The use and abuse of vegetational concepts and terms. Ecology 16:284–307
Thompson P (1989) The structure of biological theories, SUNY series in philosophy and biology, State University of New York Press, New York
Valen V, Leigh M (1991) Biotal evolution: a manifesto. Evolutionary Theory 10:1–13
Valen V (1972) Laws in biology and history: structural similarities of academic disciplines. New Lit Hist 3(2):409–419
Valen V (1973) A new evolutionary law. Evolutionary Theory 1:1–30
Valen V (1977) The Red Queen. Am Nat 111:809–810, 980
Valen V (1989) Three paradigms of evolution. Evol Theor 9:1–17
Waters CK (1994) Genes made molecular. Philos Sci 61(2):163–185
Waters CK (2000) Molecules made biological. Rev Int Philos 4(214):539–564
Waters CK (1990) Why the anti-reductionist consensus won't survive: the case of classical mendelian genetics. Philosophy of Science, Proceedings 1, 125–139
Williams MB (1970) Deducing the consequences of evolution: a mathematical model. J Theor Biol 29:343–385
Wimsatt WC (1976) Reductive explanation: a functional account philosophy of science proceedings PSA. Proceedings of the Biennial Meeting of the Philosophy of Science Association, 1974, 671–710
Wimsatt WC (1979) Reduction and reductionism. In: Asquith PD, Kyburg H (eds) Current research in the philosophy of science, pp 352–377. Philosophy of Science Association, East Lansig
Woodger JH (1952) Biology and language. Cambridge University Press, Cambridge, UK
Woodger JH, Tarski A, Floyd WF (1937) The axiomatic method in biology. The University Press, Cambridge Eng

CHAPTER 6

THE CONCEPT OF THE GENE IN CONTEMPORARY BIOLOGY: CONTINUITY OR DISSOLUTION?

JEAN GAYON

Institut d'histoire et de philosophie des sciences et des techniques (CNRS/ENS), Paris, France
gayon@noos.fr

Abstract: "Gene" is a theoretical term. Like all theoretical terms, it applies to many different domains of research. Like all theoretical terms, its meaning has dramatically changed over and over in time, and it has been defined in so many different operational ways. The problem is that the descriptive content of the various definitions of the genes that exist do not coincide. This paper provides a general evaluation of this situation. Firstly it shows that the theoretical concepts of classical genetics cannot be correlated unambiguously with the theoretical concepts of molecular genetics. In fact, there is no agreement on such simple questions as: Where are the genes? When do they exist? What are they? How many? Secondly, it provides an interpretation of why biologists continue to use the word 'gene'. Three complementary explanations are proposed: scientific communication, economical stakes, and struggle for scientific authority among biological disciplines

1. GENE: A THEORETICAL TERM

It is a common attitude among philosopher to distinguish observational and theoretical terms in scientific theories. One of the clearest formulations of this distinction can be found in Nagel (1961). The meaning of an observational is fixed through an overt observational procedure. It is independent from the statements (esp. laws) that use it. For instance the 'pressure' and the 'temperature' of a gas are observational terms. Their meaning does not depend on the experimental laws (e.g. Boyle's law) that relate them. Theoretical terms have a different status; their meaning is not fixed by observational procedures. It is fixed only indirectly in the light of the eventual uses to which a given theory may be part. "Mass" or "force" are classical examples of theoretical terms in physics. Theoretical terms are understandable in relation with a complex symbolic structure (deductive or not, this is not the issue). Therefore their meaning depends on a network of highly abstract statements in which they appear. Theoretical terms may be easily recognized through

A. Fagot-Largeault et al. (eds.), The Influence of Genetics on Contemporary Thinking, 81–95.

two characteristic features. Firstly, at a given time, they apply to many diverse areas: they are conceptual "nodes", so to say. Secondly, their meaning change in time, sometimes quite dramatically, in function of succeeding theoretical frameworks.

Post-posivitst philosophers of science have abundantly criticized this classical distinction between observational and theoretical terms, because it is unclear that pure "observational terms" have ever existed in science. Nevertheless we will take this distinction has a useful approximation. We will take for granted that at a relatively big temporal scale, there are terms the meaning of which is mostly insensitive to theory change, and other terms the meaning of which is highly sensitive to theory change.

'Gene' is a theoretical term. As all theoretical terms, it applies to many different domains of research. As all theoretical terms, its meaning has dramatically changed over and over in time. This does not mean that there has never existed an operational definition of the gene. There have been indeed many operational definitions of the gene in precise empirical contexts. But this is precisely the problem. Since the term 'gene' refers to a theoretical entity, and, indeed, to the most abstract concept of the genetical theory, it should not come as a surprise if it has been defined in so many different operational ways, such as: Mendelian unit (i.e. a hereditary unit that segregates according to the Mendelian ratios in sexually reproducing species), unit of mutation, unit of recombination, unit of function, cistron, DNA sequence coding for a protein, or – as Jacob and Monod proposed when they introduced in the 1960s the notion of "regulator gene" – a list of disjunctive possibilities. In the latter case, the definition of the gene is: a continuous DNA sequence that either codes for a protein, or plays the role of a 'regulator', or the role of an 'operator', or the role of a 'promoter'. We have mentioned here only the best-known definitions that have been used in the course of the 20th century. Other definitions have been proposed. Suffice to say that around 1980, even disjunctive definitions became problematic. As Francis Crick said in 1979 in a famous review article on split genes: 'Throughout this article I have deliberately used the word "gene" in a loose sense since at this time any precise definition would be premature' (Crick, 1979, 270, n. 1).

The problem with the definitions given above is that their descriptive content does not rigorously coincide. Even before the emergence of molecular genetics, the classical genetical definitions (segregating unit, recombination unit, mutation unit, unit of function) *did not* actually coincide (Green and Green, 1949; Pontecorvo, 1952); this is why Seymour Benzer used the notion of 'cistron' (a functional unit defined by a *cis-trans* or pseudo-allelism test; Benzer 1955, 1959). For a small number of years, it seemed that this Mendelian definition was compatible with the new molecular interpretation of genes that resulted from the discovery that the hereditary material consisted in DNA. But this situation did not last for long. Jacob and Monod's discoveries about genetic regulation revealed that the actual organization of DNA is subtle, and sometimes hardly compatible with the traditional genetical view of chromosomes as sequences of discrete Mendelian genes. Dozens of discoveries in molecular biology in the past 30 years have made doubtful that the traditional (or non molecular) concept of the gene could be translated into molecular terms with no ambiguity. Moreover, it is unlikely that a comprehensive molecular definition of the gene can be offered. Nevertheless, modern biologists go on using the term 'gene'. This paper aims at providing a general evaluation of this situation.

In the rest of this article, I raise two questions. Firstly I examine whether the classical concept of the gene can be translated into molecular terms or not. Secondly, I try to provide an encompassing picture of the gene in contemporary biology. I say "picture" rather than "concept", because the issue is not only epistemic; it is also social.

2. IS THE CONCEPT OF THE GENE TRANSLATABLE INTO MOLECULAR TERMS?

This question is part of a broader question, about the reducibility of classical genetics to molecular genetics. For most biologists, it is obvious that classical genetics has been indeed reduced to molecular biology. For the majority of philosophers who have examined this question, classical genetics has not been reduced to molecular biology and cannot be reduced to it (Hull, 1972, 1974; Kimbrough, 1979; Kitcher, 1984. A plea for reducibility can be found in Schaffner, 1960. For a comprehensive synthesis: Sarkar, 1998). Of course, the solution of the problem depends on what is meant by 'reduction'. There is an intuitive notion of theory reduction. According to it, a given scientific theory is reduced to another theory if the new theory accounts for the data explained by the former theory, but has greater scope, and produces more accurate predictions. Most scientists share this conception of reduction. Some philosophers have proposed a more or less formal elaboration of it (Kemeny and Oppenheim, 1956). If we apply this relatively tolerant notion of reduction to genetics, it is reasonable to say that classical genetics has been reduced to molecular biology: molecular biology accounts for the phenomena that classical genetics explained, it has greater scope (i.e., it explains many phenomena that classical genetics did not explain, such as the replication of the genetic material, gene action, etc.), and its predictions are more accurate.

Woodger and Nagel, among others, have proposed a stronger notion of reduction (Woodger, 1952; Nagel, 1961). According to Nagel, a theory is reduced to another one if and only if two conditions are satisfied: (1) the primitive terms of the reduced theory must be connected with the primitive terms of the reducing theories in such a way that the descriptive contents of each couple of corresponding terms has the same descriptive content; (2) the fundamental theoretical statements (or fundamental laws) of the reduced theory must be derivable from statements of the reducing theory (with the addition of appropriate initial conditions if necessary). If these two requirements are not satisfied, the older theory is not reduced by the new theory, but rather replaced by it. Philosophers such as Feyerabend and Kuhn have defended that replacement is always the case in history of science (Kuhn, 1970; Feyerabend, 1962).

Thus, in this strong sense of reduction, if classical genetics is reducible to molecular genetics, then (1) all its primitive terms (gene, dominance, mutation, recombination, etc.) must have unambiguous equivalents; (2) genetical 'laws' must be deducible from fundamental statements in molecular biology (with, if necessary, statements specifying initial conditions). In practice, no biologist has ever tried to do this and no philosopher has ever provided a full demonstration. And probably, nobody will ever do it, because nobody knows what the exact list of 'primitive terms' and 'fundamental laws' of both theories is. Most of the discussions among philosophers have been concerned only with the condition of 'connectability' of the basic vocabulary of genetics with the molecular vocabulary. Hull is probably the only author who has considered this question with some detail, considering not only the concept of the gene, but a limited list of other concepts, such as 'homozygote' and 'heterozygote', 'dominant', 'recessive', 'crossover', and 'epistasy'. In fact, most authors have discussed only the question whether the term 'gene' has a molecular equivalent. This is in fact a good strategy: if the most central concept of genetical theory is not translatable into molecular terms, then one does not need to go further: any program of reduction of classical genetics to molecular biology will fail, just because no genetical knowledge can be imagined that does not make use, directly or indirectly, of the concept of the gene.

Steven Kimbrough (1979) has provided a very elegant treatment of the question whether the concept of the gene is translatable into molecular terms. I will use freely here the general spirit of his method and apply it to data that were not available when he wrote his paper. Kimbrough says that there are three possible kinds of reduction of theoretical terms: strong type/type reduction, limited type/type reduction, token/token reduction. Basically, this categorization is itself taken from Fodor (1975).

Let us call Z, Y, X, . . . , the basic vocabulary of the reduced theory (classical genetics), and A, B, C, . . . , the basic vocabulary of the reducing theory (molecular biology).

Strong type/type reduction means a one-to-one correspondence between two theoretical terms:

$$X = A$$ [1]

In the case of the gene, this means that, for any particular instance X_i of the predicate 'gene', we will be able to establish a correspondence with a particular instance A_j of a single molecular predicate. As early as 1960, Kenneth Schaffner proposed that 'DNA sequence' was the appropriate candidate. Strong type/type reduction for the predicate 'gene' will be satisfied only and only if we can say that 'a given gene' is 'a given DNA sequence':

$$[\text{Gene}]_i = [\text{DNA sequence}]_j$$

[1] Or, at least, $X \rightarrow B$.

This formula means that the terms ‘gene’ and ‘DNA sequence’ have the same reference though they have not the same meaning (their meaning is fixed through their proper theoretical context).

Limited type/type reduction is less requiring. It admits a one-to-many relation between the reduced theory and the reducing theory:

$$X = A \vee B \vee C \vee \ldots$$ [2]

For instance:

$[\text{Gene}]_i = [\text{a continuous and discrete DNA sequence coding for a protein}]_j$

or : $[\text{Gene}]_i = [\text{a continuous and discrete DNA sequence transcribed into a tRNA}]_k$

or : $[\text{Gene}]_i = [\text{a continuous and discrete DNA sequence acting as an operator}]_l$

or : $[\text{Gene}]_i = [\text{a continuous and discrete DNA sequence acting as a promoter}]_m$

or : $[\text{Gene}]_i = \ldots.$

or : $[\text{Gene}]_i = \ldots.$

This situation is acceptable for theory reduction, because it allows to unequivocally infer from a molecular structure to a ‘classical gene’. The reverse situation (many-to-one) is not acceptable. If a given molecular structure could correspond to different genetical entities, then it would be impossible to deduce the laws of the reduced theory (genetics) from the reducing theory (molecular biology). Genes and genetical processes may supervene upon molecular entities and processes but not the reverse.

If type/type reduction is not possible, a third possible kind of reduction remains. *Token/token reduction* requires that any given instance of the term to be reduced be identified with a particular description in the language of the reducing theory. In the case of genes, this means that there will always be a particular molecular description of a given gene. Thus it will be possible to translate a given genetical situation in a particular situation, but this will no guarantee that the gene as a theoretical concept has a definite counterpart in the theoretical language of molecular biology.

Let us now apply our criteria. Strong type/type reduction of the predicate ‘gene’ to a molecular predicate is obviously not possible. There is no single molecular concept that corresponds to the concept of the gene in classical genetics. Perhaps a gene is a DNA sequence, but very few DNA sequences can be truly said to correspond to genes. It is not true that the term ‘gene’ has the same reference as the term ‘DNA sequence’. Limited type-to-type reduction is not possible either,

[2] Or, at least: $X \rightarrow (A \vee B \vee C \vee \ldots)$.

because there are plenty of phenomena in molecular biology that are ambiguous from a Mendelian point of view. Here are a few examples among many.

In prokaryotic cells, it became clear in the 1970s that the operator, promotor and coding sequences of the lactose operon (and many other similar structures) were not discrete adjacent structures, as Jacob and Monod believed in the early 1960s, but rather overlapping structures. As we know it today, the fine structure of the regulatory sequences of the lactose operon is incredibly more complex than anything that could be anticipated in the 1960s. In his retrospective book *The Operon*, Benno Müller-Hill explains that there are indeed three *lac* operators: two of them (O_3 and O_1) are inserted within the *lac* promoter region, the last one (O_2) being inserted within the structural gene coding for beta-galactosidase. Furthermore, the promoter region also includes a binding site for a protein named CAP, which interacts with RNA polymerase and activates transcription. Therefore, what Jacob Monod believed to be discrete adjacent DNA sequences, the 'promoter gene', the 'operator gene' and the 'structural genes' overlap in many different and subtle ways (see Figure 1). This means that given DNA sequences may behave as several Mendelian 'genes'. This is typically a many-to-one situation, where entities of the reducing theory (molecular structures) correspond to several different entities in the reduced theory (several genes).

In Eukaryotic cells, the situation is worse. Since the late 1970, an impressive list of phenomena implying that a given DNA sequence may behave as several Mendelian genes has been discovered: alternative splicing (different mRNAs made from different combinations of exons in split genes[3]), *trans*-splicing (spicing together of exons from transcripts encoded by separate genes), direction of transcription (DNA been transcribed in both direction), frame shift (there is a shift by one or two bases when transcribing DNA, leading to two or three different definitions of the triplets that code for amino acids), use of different genetic codes in the same cell (mitochondria and nucleus), RNA editing (post-transcriptional mRNA modifications), repeated genes (the same classical gene correspond to different positions or *loci* on one or several chromosomes), nested genes (genes that reside within an intron of a second gene), use of different strands of DNA for the making of one single protein, assembled genes (somatic coding sequences made of different DNA germinal 'genes'), etc. Some of these phenomena are rare. This is the case of frame shift, which is observed mainly in viruses. Others are extremely common in all eukaryotic organisms (alternative splicing). For a very complete discussion of these phenomena, and their bearing on the question of the unity of the gene concept, see the remarkable review by Portin, 1993 (see also Sarkar, 1998, 156–159).

[3] That is genes made of exons (coding sequences) and introns (or 'intervening sequences'). These 'genes-in-pieces' are first transcribed into a primary transcript. Then the non-coding sequences of this mRNA are excised. Finally, the remaining coding sequences are spiced, leading to a mature mRNA ready to be translated into a polypeptidic sequence. If a gene is subject to alternative splicing, different combinations of exons are possible, leading to different proteins.

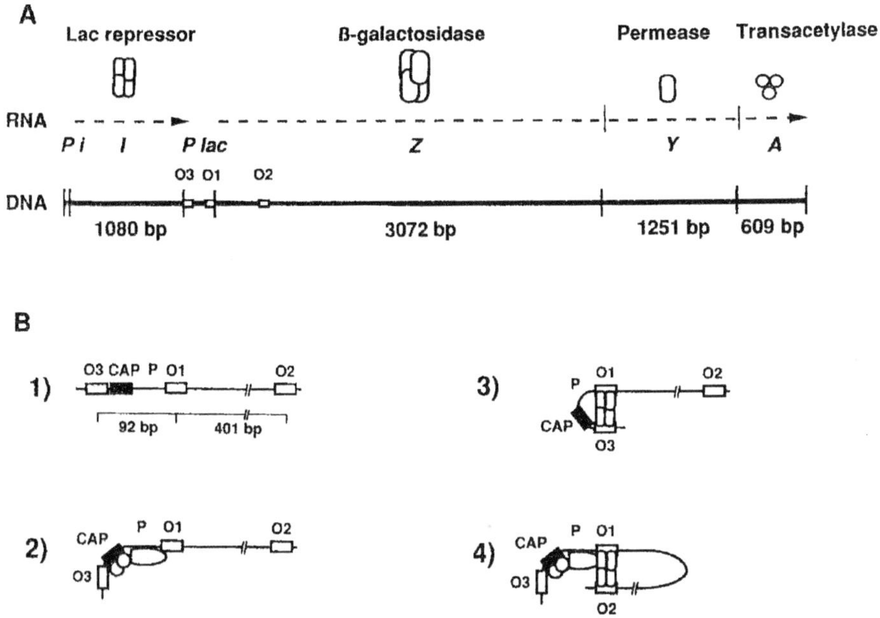

Figure 1. Modern model of repression in the lactose operon. A. The *lac* system. B. 1. The *lac promoter* region. 2. Unrepressed transcription. 3. Repression by tetrameric Lac repressor binding to O_1 and O_3. The CAP site and the promoter are depicted as unoccupied. 4. Repression by tetrameric Lac repressor binding to O_1 and O_2. The CAP protein and the RNA polymerase are (hypothetically) represented as occupied. (from Benno Müller Hill 1996, 166). This figure illustrates the tight overlapping of several regulating sites (operator sites and CAP receptor site) with the promoter site and with one of the coding sequence

Each of these phenomena reinforces the same idea: molecular structures are massively ambiguous in terms of their correspondence with classical genes. Case by case molecular descriptions of classical genes *are* always possible. But it is not possible to *a priori* infer from a given molecular structure to a unique Mendelian gene. Of course, one could say that the term 'gene' is equivalent to a very long disjunctive list of molecular situations. But, since this list would be extremely long, and cannot be exhaustively given, the only sound solution is that classical genetics as a *theory* has not been reduced by molecular biology, but rather *replaced* by molecular biology. Molecular biology accounts for phenomena that are explained by classical genetics and it also explains many phenomena that were not explained by classical genetics. Therefore molecular genetics is a more powerful theory (if the word 'theory' has still a sense – a rather difficult issue that I will not discuss here). But it can hardly be said that the theoretical concepts of classical genetics can unambiguously be correlated with the theoretical concepts of molecular biology, especially in the case of the gene.

3. WHAT IS A GENE IN CONTEMPORARY BIOLOGY? A COMPREHENSIVE PICTURE

In practice, molecular biologists, and biologists in general, go on using the term 'gene'. They behave as if the coexistence of the Mendelian vocabulary and the molecular vocabulary were not a problem. The question, then, is whether a unified concept of the gene can be constructed or not. If yes, what does it consist in? If not, why is the term 'gene' still used by biologists.

Strikingly, very few modern biologists have faced the problem of the coherence of the concept of gene. Defining the concept of the gene has been a major issue for classical genetics, from the very beginning of this science to the ultra sophisticated approach of Benzer in the 1950s. But it has never been a serious issue for molecular biologists. Almost all the papers on the subject have been written either by historians or philosophers of science, or by geneticists with a distinctive interest for the history of their discipline (Falk, 1994; Carlson, 1991; Portin, 1993).

Two prominent molecular biologists, Singer and Berg, have proposed a specifically molecular definition of the gene:

> a gene is a combination of DNA segments that together constitute an expressible unit, expression leading to the formation of one or more specific functional gene products that ay either be RNA molecules or polypeptides. The segments of a gene include (1) the transcribed region (the transcription unit), which encompasses the coding sequences, intervening sequences, any 5' leader and 3' trailer sequences that surrounded the ends of the coding sequences, and any regulatory segments included in the transcription unit, and (2) the regulatory sequences that flank the transcription unit and are required for specific expression."
>
> (Singer and Berg, 1991, 622)

Though this definition is constructed in reference to eukaryotic organisms (with 'split genes'), it encompasses a very large array of the objects that contemporary molecular biologists name 'genes'. Nevertheless, it has two weaknesses. Firstly, it leaves out the delicate problem of enhancers. Enhancers are regulatory sequences, generally small, that are located sometimes very far from the 'genes' which they control. Quite often too, they control many different 'genes'. Are these enhancers parts of the gene(s) that they control, or are they themselves 'genes'? The other problem is that Singer's and Berg's definition also ignores the fact that the classical or 'genetical' concept of the gene is still widely used in many biological disciplines (such as evolutionary biology, human genetics, medical genetics...). But Singer and Berg's definition conflict with the classical definition of the gene as a cistron. From the classical point of view of classical genetics, the gene as Singer and Berg conceive it is in fact *several* genes. We are thus backing to the problem examined in the previous section. The modern use of the gene concept involves a tremendous number of one-many and many-one relations between classical genes and molecular equivalents.

In his book *Les Secrets du gène*, François Gros has a section entitled 'What is a gene?'. He acknowledges that defining the concept of the gene has become today 'a very delicate problem', and he does not try to provide a solution to that problem. I would like here to explain why defining the gene is now such a 'delicate problem'. Since genes are so important in our modern worldview, we may ask a few simple questions about them, questions that can be formulated with some of the most universal cognitive categories: When? What? When and how? How many?

Where? Where are the genes? This question seems trivial. Aren't the genes certain pieces of DNA, located in each cell of each organism? In fact the question is not trivial. Immunogenetics provides a fascinating illustration of the problem. Thanks to a process of somatic recombination of a small number of germinal genes, higher vertebrates are able to make several billions of different somatic genes that code for the light and for the heavy chains of immunoglobulins. Leder (1982) provides the following estimation: through the shuffling of 300 germinal DNA segments, 18 billions of somatic genes can be produced that lead to 18 billions of antibodies at least. Additional processes of somatic mutation make this number an underestimation. A striking aspect of the immunogeneticists' discourse is that they speak of 'genes' both for the germinal DNA segments that are recombined and mutated in somatic cells, and for the terminal DNA product which is actually used for the making of antibodies. Sometimes, the word 'gene' is used for the germinal DNA segments that are recombined in somatic cells:

> Tonegawa and Hozumi were able to show that the arrangement of light-chain genes is different in embryonic cells and antibody-producing cells. The activation of the genes in the course of development is accompanied by their somatic recombination: the genes are shuffled.
>
> (Leder, 1982, 75)

In the same paper, however, the same author seems to reserve the word 'gene' to the final somatic DNA product:

> The genes ultimately specifying the structure of each antibody are not present as such in germ cells (the male sperm and the female egg) or in the cells of the early embryo. Rather than harboring a set of complete and active antibody genes, these cells contain bits and pieces of the genes: a kit of components. The components are shuffled in the cells of the immune system called B lymphocytes as those cells develop and mature (...). The result is that in the mature descendants of each line a unique gene is assembled, whose information is expressed in the form of a unique antibody.
>
> (Leder, 1982, 72)

Tonegawa, who discovered this fascinating phenomenon, uses an even more striking formula:

> In the genome of a germ-line cell, the genetic information for an immuglobulin polypeptidic chain is contained in multiple gene segments scattered along the chromosome. During the development of bone-marrow lymphocytes, these gene segments are assembled by recombination *which leads to the formation of a complete gene.*
>
> (Tonegawa, 1983, 575, emphasis added)

We can observe here a characteristic hesitation of modern molecular biologists on the issue where the genes really are in an organism. Immunogeneticists admit that somatic antibodies genes are *generated* through a process of recombination of germinal genes: 'two recombinations are necessary to generate the heavy-chain gene' (Sakano et al., 1980, 676). Taken literally, this mode of expression implies that we – humans and more generally higher vertebrates – have many more somatic genes in our body than we have germinal genes. Such a situation would have been unintelligible for geneticists of the first half of the 20th century. For them, the word 'gene' had to be restricted to the hereditary material. Modern molecular genetics is largely decoupled from 'heredity'. This is a major conceptual event in contemporary biology, which certainly constitutes a major source of confusion for anyone who wants to clarify the concept of the gene.

What? What are genes made of, and what is their structure? Here again, the answer seems trivial: 'genes are made of linear sequences of nucleotides'. But for contemporary molecular biology, this formula is a rather unsatisfying approximation. Is a gene just a coding sequence, or at least a discontinuous series of exons and introns, or the (variable) association of these exons with their promoter, internal or external regulating sites, and enhancers (which are often very distant from the coding sequences? As Gros says (1986), the question of the spatial limits of a given gene is one of the biggest difficulties raised by a molecular definition of the gene. Assembled genes, nested genes and other strange phenomena increase the difficulty of saying precisely what the components and the structure of a given gene are.

When? When can something be said to be a gene in the life of a cell or of an organism? Many processes are now known that force the geneticist to recognize that a given gene is not permanently here on a chromosome, but is often the result of a developmental process. I have already evoked the case of Ig genes, which are generated by way of somatic recombination and somatic mutation of germinal genes or parts of them. Other mechanisms exist that lead contemporary molecular biologists to think of genes not only in spatial terms, but also with respect to time: splicing, all sorts of 'gene' editing mechanisms (pre- and post-transcriptional, not to speak of post-translation modification of proteins), discontinuous transcription, etc. Today, many geneticists endorse Pontecorvo's prophetic view according to which a gene should be thought of in terms of a *process* rather than a definite and stable structure:

> the genes, as unit of physiological action (...) are obviously not mega molecules. They are processes, or functions, not atomic edifices. (...) For convenience of speech we may continue to call gene the structure, when there is no danger of confusion, provided we keep in mind that we are using a figure of speech as when in French we use the same word for tongue and language.
>
> (Pontecorvo, 1952, 134)

Though the context within which Pontecorvo formulated this idea is very different from today's context (this was one year before the discovery of the double helix!), it fits admirably with the present state of molecular biology. Now, if genes are processes rather than stable structures, the temporal limits of genes (or at least of many eukaryotic and viral genes) are as much problematic as their spatial limits. Therefore the question 'What is the structure of gene' becomes less important than 'when and how something comes to play the role of a gene in the development and the metabolic life of an organism?'(Portin, 1993).

How many? Here again, modern biologists are in an uncomfortable situation. For instance, do humans have approximately 30,000 or 40,000 genes, according to the conventional definition of the gene by the sequencing consortium (a definition close to Singer's and Berg's definition quoted above), or ten times more, if we consider that eukaryotic cells contain on the average approximately 500,000 different kinds of proteins, or several billions, if we take into account – as immunogeneticists actually do, the Ig mature genes in a human organism?

All these ambiguities indicate that the theoretical framework of modern molecular genetics is really different from the theoretical framework of classical genetics. It is indeed so much different that it is highly questionable whether the gene – a concept that belongs fundamentally to *genetics*, not molecular biology– is still a theoretical concept, or just a pragmatically useful word, that has no real coherence from a theoretical point of view.

My conviction is that the concept of the gene is no longer a theoretical concept as it was in the heydays of classical genetics, or even of classical molecular genetics (in the 1960s). There is probably no hope to construct a general or unified concept of the gene. At best, such a concept would be an indefinite disjunctive list of many possible structures and processes. There was a very short period of time, in the 1960s, when the gene was a genuine molecular concept, because it was still then possible to translate the most advanced classical concept of the gene (Benzer's cistron) into a non-ambiguous molecular formula. This period corresponds to the time when Jacques Monod and François Jacob exposed in a successive series of legendary papers that genes, from a molecular point of view, could be defined as discrete adjacent segments of DNA that could have several possible functions: 'structural' (that is, coding for a protein, either cytoplasmic or a repressor), 'operator', 'promotor'. Even though these distinctions were not absolutely clear in the early sixties (especially because Monod and Jacob did not think at the beginning that the repressor gene coded for a protein)[4], it is retrospectively possible to say that during a small number of years, molecular biologists could reasonably think that the genetical concept of the gene could be translated into a non ambiguous disjunctive list of discrete molecular structures with definite functions. But this theoretical

---

[4] Here is Jacob's and Monod's original definition of the 'structural gene hypothesis': 'the DNA message contained within a gene is both necessary and sufficient to define the structure of a protein' (Jacob and Monod, 1959, 1961, 318). Since Jacob and Monod did not believe that the repressor was a protein when they proposed the *lac* operon hypothesis, it was natural for them to make a distinction between 'structural gene' and 'regulator gene' – that is a gene 'controlling' a repressor molecule (Jacob et al., 1960). When it was discovered that the repressor was a protein, this distinction became a source of confusion. Another cause of confusion was the introduction of the notions of 'operator' and 'promoter'. In their first characterization of the operator, Jacob and Monod did not characterize this DNA sequence as a 'gene' (Jacob et al., 1960). But a year later, they spoke of an 'operator *locus*', whose function is to bind with the repressor molecule (Jacob and Monod, 1961). In this paper, the authors give a diagram which explicitely contains the terms 'regulator gene', 'operator gene' and 'structural gene'. Finally, in 1964, Jacob and Monod introduced a fourth category of molecular gene, the 'promotor' (Jacob and Monod, 1964).

situation did not last very long. Since the 1970s, it has become increasingly clear that modern 'genes', as molecular biologists see them are hybrid conceptual constructions, which would be more rigorously designated without using the term 'gene'. As Petter Portin lucidly said a few years ago:

> Our knowledge of the structure and function of the genetic material has outgrown the terminology traditionally used to describe it. It is arguable that the old term gene, essential at an earlier stage of the analysis, is no longer useful, except as a handy and versatile expression, the meaning of which is determined by the context. In that case, the challenge is to devise a new terminology for use when precision is needed.
>
> (Portin, 1993)

Why is it, then, that contemporary biologists persist in using the term 'gene'? I think that three different reasons account for this situation (Gayon, 2004).

The first reason is that scientific communities, as all human groups, need words that prove efficient for the practical purpose of communicating. Quite often, in science as in daily life, approximate terms are more useful that rigorously defined terms, because individuals always belong to various different communities, each one with its own needs and conventions. With its own ambiguities, the word 'gene' facilitates reciprocal understanding between different categories of biologists: molecular biologists, biochemists, evolutionary biologists, human geneticists, medical geneticists, and many other biological sub-disciplines. The first reason, then, for the persistence of the word 'gene' is pragmatic. The vocabulary of the gene is still useful for the cohesion of biology as such. Molecular biologists alone might (perhaps) be happy not using any longer the term and the concept of the- gene, but other categories of biologists need crucially this concept. This is especially the case of evolutionary biologists, and of all the biological disciplines for which evolution is crucial (e.g. ethology, population biology, human genetics...). For these disciplines, it is crucial that the genome be fragmented in relatively discrete units, whatever the molecular mechanisms by which this discreteness is attained. Richard Dawkins has given the most vivid formulation of this idea. For him, a gene is "any stretch of DNA, beginning and ending at arbitrarily chosen points on the chromosome", that "[competes] with allelomorphic stretches for the region of chromosome concerned" (Dawkins, 1982, 87). Or, to say it in a cruder way, genes are just the most fundamental replicators or "survival machines" (Dawkins, 1976). According to this 'evolutionary gene concept', what is important for the evolutionary biologist is the inheritance of DNA sequences as such, whatever the complexities of the gene-phenotype relationship.[5] Though not all evolutionary biologists agree with Dawkins' genocentric view of evolution, most of them would certainly follow him in his plea for an evolutionary conception of the gene, a conception that is certainly closer to the classical or premolecular conception of the gene than it is close

[5] An interesting critical discussion of the evolutionary gene concept can be found in Sterelny and Griffiths 1999, and Stotz et al., 2004.

to the molecular conception. Similar observations could be made about other disciplines for which genes are primarily thought of in terms of something that is transmitted from generation to generation (medical genetics, human genetics, behavioral genetics, etc.). Therefore, the biological community as a whole has serious reasons to keep the word 'gene' as a sort of conceptual currency, which, in spite of its ambiguities, is useful from the point of view of interdisciplinary collaboration.

A second reason for the persistence of the word gene is economical. Today, genes are not only physiological resources for the organisms. They also constitute promising technical resources, and, as such, they are virtual sources of benefits. There are indeed powerful economic reasons for keeping the word 'gene', even though the operational definitions that are used for this purpose are not theoretically satisfying. In practice, 'genes' (understood as coding sequences with their adjacent regulatory sequences) are patented. Therefore, genes are not only objects of scientific dispute; they also exist as symbolic and material resources in the social world. In the beginning of the 20th century, eugenics was a major cause of the development of genetic research in many countries. In this context, genes were important ideological resources. A century later, external factors remain as important, but they are not the same. What is socially important about genes today is that they designate important economic resources.

The last reason for keeping the word 'gene' is political. It has something to do with the relations of power that develop within the scientific communities. This point has been nicely discussed by the philosopher Michael Dietrich in a conference held at the Paris Academy of science on the occasion of the hundredth anniversary of the rediscovery of Mendel's laws (Dietrich, 2000). Michael Dietrich says that the omnipresence of genes in the contemporary biological discourse should be interpreted in the light of "the struggle for scientific authority". Even if it not possible to construct an abstract and unified concept of the gene, it is indeed important for some categories of biologists to explain the other biologists what genes really are. Since all biologists use the term, it is important for this or that biological particular community to control the use of such a powerful word. Molecular biologists and evolutionary biologists are undoubtedly the most active competitors in this struggle for scientific authority. I do not say here that this is the only reason for keeping the word "gene". But this sociological consideration is certainly a powerful factor that accounts for the persistence of the concept, especially in molecular biology (which, from a theoretical perspective, has literally dissolved the traditional concept of the gene). No contemporary biological discipline is able to offer a theoretically unified concept of the gene, but biological disciplines are engaged in a struggle for scientific authority on *the question of what it is important to know about the genes*, and on how to speak about them. Even molecular biology could easily forget the word 'gene', it is important for this dominating discipline, often presented as the new common language of all the biological disciplines, to be able to tell them what a gene is "in reality".

4. ACKNOWLEDGMENTS

I am grateful to Gabriel Gachelin and Michèle Roa (Pasteur Institute, Paris) for their generous bibliographical and scientific help. I thank Mattehw Cobb for his linguistic corrections.

REFERENCES

Benzer S (1955) Fine structure of a genetical region in bacteriophage. Proceedings of the National Academy of Sciences of USA 41:344–354.

Benzer S (1959) On the topology of the genetic fine structure. Proceedings of the National Academy of Sciences of USA 47:1607–1620.

Carlson EA (1991) Defining the gene: an evolving concept. Am J Hum Genet 49:475–487.

Crick F (1979) Split genes and RNA splicing. Science 204:264–270.

Dawkins R (1976) The selfish gene. Oxford University Press, Oxford.

Dawkins R (1982) The extended phenotype. The long reach of the gene. Freeman, San Francisco.

Dietrich MR (2000) The problem of the gene. In: Gros F, Gayon J, Morange M, Veuille M (eds) Comtes Rendus de l'Académie des Sciences de Paris, Sciences de la vie, Tome 323(12), Special issue 1900 Rediscovery of Mendel's laws, pp 1139–1146.

Falk R (1984) The gene in search of an identity. Hum Gene 68:195–204.

Feyerabend PK (1962) Explanation, reduction, and empiricism. Minnesota Stud. Phil. Sc. 3:28–97.

Fodor JA (1975) The language of thought. Thomas Y. Cromwell Company, New York.

Gayon J (1999) La génétique mendélienne a-t-elle été réduite par la biologie moléculaire? Biofutur 189, Mai:12.

Gayon J (2004) La génétique est-elle encore une discipline? Médecine/Sciences 2 (Février):248–252.

Green MM, Green KC (1949) Crossing over between alleles of the lozenge locus in *Drosophila melanogaster*. Proceedings of the National Academy of Sciences of USA 35:586–591.

Gros F (1986) Les Secrets du gène. Odile Jacob, Paris.

Hull D (1972) Reduction in genetics – biology or philosophy? Philos Sci 39:491–499.

Hull D (1974) Philosophy of biological science. Prentice-Hall, Englewood Cliffs, NJ.

Jacob F, Monod J (1959). Gènes de structure et genes de regulation dans la synthèse des proteins. C. R. Acad. Sci. Paris 249:1282–1284.

Jacob F, Perrin D, Sanchez C, Monod J (1960) L'opéron: groupe de gènes à expression coordonnée par un opérateur. Comptes Rendus des Séances de l'Académie des. Sciences 250:1727–1729.

Jacob F, Monod J (1961) Genetic regulatory mechanisms in the synthesis of protein. J Mol biol 3:318–356.

Jacob F, Ullman A, Monod J (1964) Le promoteur, élément génétique nécessaire à l'expression d'un opéron. Comptes Rendus des Séances de l'A1cadémie des. Sciences 258:3125–3128.

Kemeny JG, Oppenheim P (1956) On reduction. Philos Stud 7:6–17.

Kimbrough SO (1979) On the reduction of genetics to molecular biology. Philos Sci 46:389–406.

Kitcher P (1984) (1953) And all that. A tale of two sciences. Philos Rev 93:335–373.

Kuhn T (1970) The structure of scientific revolutions. The University of Chicago Press, Chicago.

Leder P (1982) The genetics of antibody diversity. Sc. Amer. 246 (May):72–83.

Müller Hill B (1996) The *Lac* operon. A Short History of a Genetic Paradigm. Berlin & New York, Walter de Gruyter.

Nagel E (1961) The structure of science. Harcourt, Brace & World, New York.

Pontecorvo G (1952) The genetic formulation of gene structure and action. Advances in Enzymology 13:121–149.

Portin P (1993) The concept of the gene: short history and present stats. The Q Rev Biol 68:173–223.

Sakano H, Maki R, Kurosawa Y, Roeder W, Tonegawa S (1980) Two types of somatic recombination are necessary for the generation of complete immunoglobulin heavy-chain genes. Nature 286 (14 August):676–683.

Sarkar S (1998) Genetics and reductionism. Cambridge University Press, Cambridge.
Singer M, Berg P (1969) Genes and genomes: a hanging perspective. University Science, Mill Valley.
Schaffner KF (1969) The Watson-Crick model and reductionism. Br J Philos Sci 20:325–348.
Sterelny K, Griffiths PE (1999) Sex and death: an introduction to the philosophy of biology. Chicago University Press, Chicago.
Stotz K, Griffiths PE, Knight R (2004) How biologists conceptualize genes: an empirical study. Studies in History and Philosophy of Biomedical Sciences 35:647–673.
Tonegawa S (1983) Somatic generation of antibody diversity. Nature 302 (14 April):575–581.
Woodger JH (1952) Biology and language. Cambridge University Press, Cambridge.

CHAPTER 7

THE INFLUENCE OF GENETICS ON PHILOSOPHY OF SCIENCE

Classical genetics and the structuralist view of theories

PABLO LORENZANO

Universidad Nacional de Quilmes – Consejo Nacional de Investigaciones Científicas y Técnicas, Quilmes, Argentina, pablol@unq.edu.ar

Abstract: Taking as starting point the first textbook of classical genetics that clearly exemplifies all the features that Kuhn takes to be constitutive of a science textbook, Sinnott and Dunn's (1925), as well as Darden's (1991) and Schaffner's (1980, 1986, 1993) analyses of the structure of biomedical and/or biological theories, I will discuss the problem of the existence of laws in biology. The framework of this discussion is provided by the structuralist conception of theories. The result of this analysis will be the identification of the fundamental law of classical genetics: the law of matching, which satisfies all weak necessary conditions for law-likeness that are postulated by the structuralist approach of theories, and the recognition of the so-called 'Mendel's Laws' as special laws of classical genetics. This shows that the structuralist view is capable of providing an interesting perspective on genetics, which, in turn, has a positive influence on philosophy of science, because it shows we have a framework at hand in which important philosophical problems can fruitfully be addressed.

1. INTRODUCTION

For a long time, the philosophical analysis of science was generally limited to physics, neglecting other fields of science such as biology. However, over the last decades, philosophy of biology has become one of the most active and promising branches of philosophy of science. Nevertheless, most of the debates over the theoretical structure of biology either took no notice of important changes in general philosophy of science, or have focused almost entirely on evolutionary biology.[1]

---

[1] Beatty (1980, 1995), Brandon (1981), Cadevall i Soler (1988), Ereshefsky (1991), Gould (2002), Hull (1974), Kitcher (1984, 1989), Lewontin (1974), Lloyd (1988), Mayr (1991a, b), Moya (1989),

A. Fagot-Largeault et al. (eds.), The Influence of Genetics on Contemporary Thinking, 97–113.

Genetics, in spite of being one of the most recent biological theories, is undoubtedly one of the most important scientific theories because of the role it plays in clarifying the riddle of heredity and its enormous relevance to technology. So it comes as little surprise that genetics attracts the attention of ever more philosophers.[2]

Taking as starting point the first textbook of classical genetics that clearly exemplifies all the features that Kuhn takes to be constitutive of a science textbook, Sinnott and Dunn's (1925), as well as Darden's (1991) and Schaffner's (1980, 1986, 1993) analyses of the structure of biomedical and/or biological theories, I will discuss the problem of the existence of laws in biology. The framework of this discussion is provided by the structuralist conception of theories. After analyzing the structuralist notion of a fundamental law, I will identify the law of matching as the fundamental law of genetics and shown how the special laws of the theory, in particular Mendel's first and second law (the law of segregation and the law of independent assortment), can be obtained from it. In doing so I show that genetics possesses all characteristics that have been considered to be essential features of empirical theories by the structuralist approach, including that of possessing a fundamental law. Thus, genetics constitutes a successful application of the structuralist theory of science, providing an argument in favor of that very philosophical point of view.

2. TEXTBOOKS AND CLASSICAL GENETICS

Ludwik Fleck (1935) and Thomas Kuhn (1959, 1962/1970) have indicated the important role that textbooks play in science teaching. But, though Fleck is the first one admitting that "the initiation to the science is realized in agreement by pedagogic special methods" (Fleck, 1935, 148) through the textbooks, it is Kuhn, who, following Fleck, indicates that "[t]he single most striking feature of this education is that, to an extent totally unknown in other creative fields, it is conducted entirely through textbooks" (Kuhn, 1959, 228) and provides the first analysis of them. In this analysis, Kuhn introduces for the first time the concept of a "paradigm": "these books exhibit concrete problem solutions that the profession has come to accept as paradigms, and they then ask the student, either with a pencil and paper or in the laboratory, to solve for himself problems very closely in both method and substance to those through which the textbook or the accompanying lecture has led him. Nothing could be better calculated to produce 'mental dispositions' or *Einstellungen*" (Kuhn, 1959, 229). It is through the acquaintance with

Rosenberg (1985), Ruse (1973), Schaffner (1993), Sintonen (1991), Sober (1984, 1993), Thompson (1989), Tuomi (1981), Tuomi and Haukioja (1979), Wassermann (1981) and Williams (1970) are some of the authors that have been working on the clarification of the structure of population genetics and of the theory of evolution by natural selection.

[2] Attempts to clarify classical and molecular genetics have been made by Beurton et al. (2000), Dawe (1982), Hull (1974), Keller (2000), Kitcher (1984, 1993), Kyburg (1968), Lindenmayer and Simon (1980), Morange (2001), Rizzotti and Zanardo (1976), Ruse (1973), Sarkar (1998), Schaffner (1993) and Woodger (1959).

the standardized, accepted and shared examples, the paradigms (which Kuhn later called "exemplars" (1962/1970)), which are used in the attempt to solve problems raised in the textbook or by the teacher at the classroom that the students learn, in a not discursive way, to see "[w]hat are the fundamental entities of which the universe is composed", "[h]ow do these interact with each other", "[w]hat questions may legitimately be asked about such entities and what techniques employed in seeking solutions", and to solve new problems or "puzzles", similar to the previous ones, which arise in the course of the scientific investigation (or "normal science") (Kuhn, 1962/1970, 4–5). Besides, "[v]erbal definitions [...] have little scientific content when considered by themselves. They are not full logical specifications of meaning (if there are such), but more nearly pedagogic aids. The scientific concepts to which they point gain full significance only when are related, within a text or other systematic presentation, to other scientific concepts, to manipulative procedures, and to paradigm applications" (Kuhn, 1962/1970, 142). "[T]extbooks treat the various experiments, concepts, laws, and theories of the normal science as separately and nearly seriatim as possible" (Kuhn, 1962/1970, 140), which provides a distorted image of the history of the discipline. In fact, only few historical references appear in textbooks "either in an introductory chapter or, more often, in scattered references to the great heroes of an earlier age. Yet the textbook-derived tradition in which scientists come to sense their participation is one that, in fact, never existed" (Kuhn, 1962/1970,137–138).

The book by Sinnott and Dunn is the first textbook of classical genetics that clearly possess all the features that Kuhn identifies in a science textbook. This book was designed as an introduction to genetics, aiming "to set forth the essential principles of genetics in as clear and concise a manner as possible" (xvii) and to be used in elementary college courses. It first characterizes genetics in general terms in the initial chapters ("Chapter I **The Science of Genetics**" and "Chapter II **Heredity and Variation**"),[3] and then situates the discipline historically by telling what has been called the "traditional account" (Olby, 1979), the "orthodox image" (Bowler, 1989) or the "official story" (Lorenzano, 1995),[4] which basically presents its history in a continuous, accumulative and linear way, thereby assigning Mendel

[3] "That branch of the science of biology which is concerned with the phenomena of inheritance and variation and which particularly endeavors to discover the laws governing these similarities and differences between individuals related to one another by descent is called *genetics*" (Sinnott and Dunn, 1925, 6)."The chief aim of genetics is to discover, to classify, and to explain the facts of heredity and variation. Heredity is the tendency of animals and plants to resemble their ancestors and relatives; whereas variation is the tendency to depart or differ in any particular from the other of their kind" (Sinnott and Dunn, 1925, 17).

[4] After the Argentinean film that was awarded the Oscar as best foreign film in 1985, "La historia oficial" ("The official story"; the Spanish term "historia" stands for both story and history, hence the play of words).

a central place.[5] It goes on to introduce the basic concepts such as the one of an individual (parental one or progeny), a cross between individuals, the characteristics of the individuals (or phenotype) and the factors, factor-units or genes of the individuals (or genotype) that determine the characteristics. It presents the results of hybrid crosses (basically, the phenotypic proportions 3:1 for the monohybrids, 9:3:3:1, 9:3:4, 9:7, 12:3:1, 13:3, 15:1 for the dihybrids, and 27:9:9:9:3:3:3:1 for the trihybrids, those corresponding to continuous characters and those where linkage takes place) and the "principles" or "laws" by which they are governed,[6] the most important of which are that of segregation[7] and that of independent assortment.[8] Besides all this, it contains shared examples, paradigms (or exemplars), and a series of problems to be solved by the student.[9]

3. LAWS AND EXEMPLARS IN BIOLOGICAL THEORIES

It has been indicated that in the biological and/or biomedical sciences acquaintance with a number of shared examples is very important, because they seem to lack of laws or generalizations of wide or even universal scope (Schaffner, 1986; Darden, 1991). And, even when the "laws" or "principles" are enunciated as general, they do not seem to possess the properties (or characteristics) traditionally attributed to laws, e.g. universality and necessity. In fact, the two most important arguments against the existence of *laws in biology* refer either to their *non-universality* or their *evolutionary contingency*. The first point is due to Smart (1963) and emphasizes that what we usually consider to be biological laws lack of the kind of universality that is required to be a law ("in strict sense" (Smart, 1963) or "fundamental" (Hempel and Oppenheim, 1948)) since they often make (at least implicit) reference to particular entities and do not hold without exceptions. The second point is due to Beatty (1995),[10] who argues, based on a an analysis of the concept of lawfulness in terms of nomic or natural necessity, that

---

[5] For an analysis of this historiographic position, see, among others, Lorenzano (1995, 1997, 2002b, to appear).

[6] Where "[a] law is thus a brief statement or explanation of some uniform and constant relationship which has been found to hold through a large series of natural events" (Sinnott and Dunn, 1925, 36).

[7] "The essential feature of the mechanism of segregation, therefore, lies in the circumstance that a factor carried by the gametes of one parent and its contrasting factor carried by the gametes of the other parent, come together and coexist for a generation in the cells of the resulting hybrid offspring *without blending or losing their identity*; and that when such a hybrid individual produces its own sexual cells, in turn, these two factors become completely and cleanly separated again, or *segregated* from one another, each of the new gametes being entirely pure, containing either the one factor or the other but *never both*" (Sinnott and Dunn, 1925, 51–52).

[8] "The particular combination of factors which enters the F_1 plant from each parent (round with yellow and wrinkled with green in this case) has no effect whatever upon the way in which they are associated in the gametes formed by this F_1 plant. *Their assortment is independent*" (Sinnott and Dunn, 1925, 67).

[9] "Perhaps the most novel feature of the present volume is its series of Questions for Thought and Discussion, Problems and Reference Assignments" (Sinnott and Dunn, 1925, xvii).

[10] Elaborating a thesis supported by Gould (1989).

the biological generalizations are of two types: either "they are only mathematical, physical or chemical generalizations (or deductive consequences of mathematical, physical or chemical generalizations and initial conditions)" (Beatty, 1995, 46), or "distinctively biological generalizations" (Beatty, 1995, 47). If they are generalizations of the first type, they cannot be considered biological laws; whereas if they are of the second kind, they describe contingent results of the evolution and therefore lack *natural* or *nomic necessity*, which deprives them of the status of a law of nature.[11]

This has led some to claim that there are no laws in biology; and since they regard laws as essential to theories they then conclude either that there are no theories in biology or that biological theories possess a structure that is very different from that of theories within the physical sciences. The latter position is held by Darden (1991), who regards biological theories as sets of problem-solving schemes, later instantiated in shared examples or exemplars, and by Schaffner (1980, 1986, 1993), who claims that biomedical theories are best characterized as series of overlapping models.[12]

This raises the question of whether the problem-solving schemes, shared examples, exemplars, or models that constitute a theory are unconnected or whether they are somehow related to one another. And if the latter were the case one would also like to know how they are (or have to be) related to one another in order to for them to belong to *one and the same theory*. One might respond to this question by denying that there is one particular feature (or set of features) that all elements of theory share and argue that the case of biological models is analogous to Wittgenstein's games (1953, §66 and ff.): what ties different paradigms, shared examples, exemplars, or models, together and what makes them belong to the same theory is some kind of family resemblance between them rather than the existence of a fixed set of shared features, providing necessary and sufficient conditions for theory-membership. However, this answer begs the question because we still want to know in what sense the different elements of a theory are similar to each other. It seems unlikely that the desired similarities can be read off from the mere appearance of these systems, and this is all that the Wittgensteinean can appeal to. Moreover, what matters is not that these systems are similar to each other in appearance but rather that they share certain structural features: the paradigms, shared examples, exemplars, or models that belong to the same theory possess the same structure, meaning that they can be regarded as specifications of one and the same structure or of a more general underlying scheme. In order to develop this idea and to show how it can be applied to the case of classical genetics we now introduce that structuralist notion of a fundamental law.[13]

---

[11] For a more thorough discussion of the arguments of non-universality and of evolutionary contingency see Lorenzano (2001, 2006).

[12] For a discussion of this position see Lorenzano (2002c).

[13] See Balzer et al. (1987) or Díez and Lorenzano (2002) for a detailed discussions of the structuralist approach.

4. THE STRUCTURALIST NOTION OF FUNDAMENTAL LAW

Bas van Fraassen, one of the most important figures of the semantic approach to theories to which also the structuralist conception belongs, observes that "[w]hen philosophers discuss laws of nature they speak in terms of universality and necessity" (1989, 1). The two above-mentioned arguments against the existence of biological laws refer precisely to the lack of universality and necessity of the laws of biology. However, if one accepts these criteria, it seems that one should not only discard the laws of biology, but also the allegedly more respectable laws of physics. In fact, due to the lack of unproblematic criteria for lawlikeness,[14] van Fraassen (1989) proposes that we give up the category of a law of nature altogether. His criticism of the concepts of natural law and nomic necessity (see also van Fraassen, 1977) and his skeptical attitude towards them is shared by other authors, for instance Swartz (1995). This skeptical attitude, however, does not imply that there are no fundamental equations or basic principles of theories which guide scientific practice; rather these equations are re-conceptualized not as *laws of nature* but as *scientific laws* (Swartz, 1995) or *laws of the models* (van Fraassen, 1989, 1993). Such laws are not regarded as empirical regularities governing the natural world independently of whether intelligent beings possess knowledge of their truth and necessity or of whether an appropriate symbolic representation for some of those regularities has been developed; such laws are human creations. Laws, on this view, refer to those regularities of the natural world (or better of the *modeled world*) that are known to us and that are represented in the appropriate symbolic form in a collective effort to explain, predict and control parts of the world.

However, despite sustained efforts, we still don't have a satisfactory concept of scientific law at hand, i.e. we still lack an adequate set of necessary and sufficient conditions serving as criteria for a statement to be considered a "(scientific) law".[15] Worse still, "[i]t is likely that no such set of conditions can ever be found that would appear satisfactory to everybody since the notion of a law is a strongly historical, discipline-dependent kind of notion" (Balzer et al., 1987, 15). Within the structuralist tradition, discussion of the notion of a law have usually focused on what Stegmüller (1973) called a "fundamental law of a theory".[16] And when the criteria for a statement to be a fundamental law are discussed, there is a tendency

[14] See Weinert (1995) for a discussion of the concept of law of nature.

[15] See Stegmüller (1983) and Salmon (1989) for an analysis of the difficulties which confront the classical explication of the concept of scientific law.

[16] See, for instance, Balzer (1979a), Balzer et al. (1987), Bartelborth (1988), Moulines (1978/1982, 1991), Sneed (1971), Stegmüller (1973, 1976, 1978, 1979a, 1979b, 1986). The expression "fundamental law" and the later one "special law" are not used here in the sense of Fodor (1974, 1991), i.e. in the sense of laws which belong to different kinds of sciences, the former to fundamental or basic sciences and the latter to special sciences. Rather these notions are used in the sense of the structuralist conception, i.e. as referring to different kinds of laws of one and the same theory. And as one we will see later, the expression "fundamental law" is not used in the sense of the classical explication mentioned above either.

to speak about "necessary conditions" (Stegmüller, 1986), "weak necessary conditions" (Balzer et al., 1987) or about "«symptoms», some of them even formalizables" (Moulines, 1991), even though it is admitted that "in every particular case of reconstruction of a given theory, as a general rule, it seems to be relatively easy to agree, on the basis of informal or semiformal considerations (for example, on its systematizing role or its quasi-vacuous character), that a certain statement should be considered as a fundamental law of the theory in question" (Moulines, 1991, 233).

Stegmüller (1986) mentions two criteria as necessary conditions for something to qualify as a fundamental law: first, having a cluster or synoptic character and, second, being valid in every intended application of the theory. The first criterion, having a cluster or synoptic character, which made its first appearance in the structuralist literature in Stegmüller (1979a, 1979b) and which is further discussed in Balzer et al. (1987) and in Moulines (1991) has received different formulations, some stronger than others. According to the strongest of them, "any correct formulation of the law should include necessarily all the relational terms (and implicitly also all the basic sets) and, therefore, at the end, every fundamental concept that characterize such a theory" (Moulines, 1991, 234). However, when phrased in this way, this feature, as Moulines himself recognizes (1991, 233–234), is not possessed by all possible candidates of fundamental laws; noteworthy exceptions include the fundamental laws of continuum mechanics and of electrodynamics, which, according to the analysis made by Bartelborth (1988, 19ff., 45f., 53), "do not seem to be able of reformulation as synoptic laws in a plausible and natural way" (Moulines, 1991, 234).

Weaker formulations of this criterion do not require that all the fundamental concepts occur in every fundamental law, but only that "any of the magnitudes" (Stegmüller, 1986, 23), "diverse functions" (Stegmüller, 1986, 93), "possibly many theoretical and non-theoretical concepts" (Stegmüller, 1986, 386), "almost all" (Balzer, Moulines and Sneed 1987, 19) or "at least two" (Stegmüller, 1986, 151) do. These versions are able to recognize those propositions as fundamental laws that were excluded by the stronger formulation of the criterion.[17]

The second criterion for a statement to be a fundamental law, though implicit in many structuralist writings, has been explicitly introduced by Stegmüller (1986). This criterion posits that a sentence must possess "validity in every intended application" (Stegmüller, 1986, 93). According to this criterion, it is not necessary for fundamental laws to possess an unlimited scope, to apply everywhere in space and time, or to have "one big, cosmic application", which constitutes one single or "cosmic" model (Stegmüller, 1979b; Mosterín, 1984). In fact, only the fundamental laws of some cosmological theories, which are applicable to the cosmic model, or the laws of the "Great Unified Theory" (GUT) – if such a thing exists – are universals in this sense. However, this is not the standard situation. The laws of the physical sciences, and the same goes for the biological ones, normally apply to partial and well-determined

[17] One has to bear in mind that this criterion is strongly dependent on the choice of a particular formulation of a theory because a term can be considered to be primitive, basic or fundamental only with respect to it.

empirical systems and not to the cosmic model. Moreover, this criterion allows to distinguish between fundamental laws and special laws, which, though synoptic, are only valid in some but not in every applications of the theory.

Moulines (1991) emphasizes two further aspects of lawfulness: the systematizing role of laws and their quasi-vacuous character. Fundamental laws are (empirically) quasi-vacuous in that they are highly abstract, schematic, and contain essential occurrences of T-theoretical terms in structuralist sense, meaning terms whose extensions can only be determined by presupposing the validity of the fundamental laws of the theory but which nevertheless acquire specific empirical content through a non-deductive process known as "specialization". This process, which provides most specific laws (the so-called "special laws"),[18] consists of the introduction of restrictions, or specifications to (some of the components of) the fundamental law(s), in such a way that they become progressively concrete in diverse directions until we finally obtained the so-called "terminal specializations" in which all components are specified. If the introduced specifications turn out to be the appropriate, the intended applications are said to be "successful". The quasi-vacuous character of the fundamental laws has led some authors to doubt their empirical nature and to propose to regard them as "non-empirical", "analytical", "a priori", "tautological stipulations", "mere conventions" or "mere definitions"[19]. In this vein, Moulines proposes to call this type of statements "empirically unrestricted" (Moulines, 1978/1982, 96): on the one hand they are irrefutable or empirically vacuous but on the other hand they are different from the paradigmatic examples of analytical statements such as "all bachelors are not married".

The other "symptom" mentioned by Moulines, the systematizing role of the fundamental laws, might be understood as one that makes it possible to include diverse applications within the same theory because it provides a guide to and a conceptual frame for the formulation of other laws (the so-called "special laws"), which, as we have seen above, are introduced by imposing restrictions on the fundamental laws. Due to the process of "specialization", which construes theories in a strongly hierarchic way, and the obtaining of "successful" applications, it is possible to integrate the different empirical systems, "models" or "exemplars" under the same conceptualization, in which the fundamental law(s) occupy a central place.

5. FUNDAMENTAL LAW AND SPECIAL LAWS IN CLASSICAL GENETICS

Taking into account what has been said about fundamental laws in the last section, the position we will argue for now differs both from the position of those who deny the existence of laws in biology (Smart, 1963) as well as from the position

[18] To repeat, the relation between the most general laws (or fundamental) and the most specific laws is not one of deduction, but just of specialization. For this reason the latter ones are referred to as "special laws" rather than of "derivative laws".

[19] For a discussion of the proposals of Brandon (1978, 1997), Sober (1984, 1993, 1997) and Elgin (2003) of non-empirical or a priori biological laws similar as the present one, that is, based on the structuralist conception of theories and on the case of classical genetics, see Lorenzano (2006).

of their critics (Ruse, 1970; Munson, 1975). These three authors agree that if there exist statements in genetics that could be considered as "laws in strict sense" or as "fundamental laws", these would have to be found among the so-called 'Mendel's laws'. Setting aside the questionable attribution of these laws to Mendel,[20] we do not agree with this point of view. Neither of these laws (that is, neither the law of the segregation, nor the law of the independent assortment) is sufficiently schematic and general in that all, or almost all, terms of the theory are connected to each other through these laws, or in that they are accepted by the community of geneticists as valid in every application, or in that they provide a conceptual frame in which the special laws of classical genetics can be formulated. For this reason, these laws cannot be regarded as the fundamental laws of genetics. And worse still, at least for those who claim that genetics possesses at least one fundamental law, so far no such law has emerged in the literature on genetics.[21]

Nevertheless, the structuralist reconstruction of classical genetics[22] suggests that upon closer inspection we find that there actually is a fundamental law of genetics.

Classical genetics is a theory about hereditary transmission, which studies the inheritance of diverse traits or characteristics (phenotypes) from generation to generation of individuals. It identifies numerical ratios (relative frequencies) in the distribution of these characteristics in the progeny, and postulates types and appropriate numbers of factors or genes (genotypes) possessed by individuals (parents and progeny), certain distributions of the parental genes in the progeny and certain relations between genes and characteristics to account for the identified distributions of the characteristics in the progeny. The fundamental law allows us to explain these distributions, by establishing that given two parental individuals – with a certain phenotype and a certain genotype along with a certain relation between phenotype and genotype –, that cross and produce progeny – which possesses a certain phenotype and a certain genotype along with a certain relation between phenotype and genotype –, a certain match[23] occurs between the distributions (relative frequencies) of the phenotypic characteristics and the distributions (expected or theoretical probabilities) of genes that are theoretically postulated. For lack of a better name, we will call this law the "law of matching". This law, though not explicitly formulated in the genetics literature, is implicit in the usual formulations of this theory, systematizing it, making sense of geneticists' practice, and unifying the problem-solving schemes, paradigms, shared examples, exemplars or models under one and the same theory. The problem-solving schemes, paradigms, shared examples, exemplars or models can be conceived as structures of the following type <*J, P, G, APP, MAT, DIST, DET, COMB*> where *J* represents the set of individuals (parental ones and progeny), *P* the set of characteristics (or

---

[20] See Bennett (1964, 1965), Lorenzano (1995, 1997, 2002b, 2006) and Olby (1979).

[21] This point is uncontroversial; see Kitcher (1984) and Darden (1996).

[22] Balzer and Dawe (1990), Balzer and Lorenzano (1997) and Lorenzano (1995, 2000, 2002a).

[23] Genetics, as virtually all empirical sciences, contains certain approximations. If these are ignored, then the match is exact. If these approximations are taken into account, then the fit is only approximate, but in a way that the distances between the coefficients that represent a theoretical distribution and those of the relative frequencies do not exceed a given ε.

phenotype), *G* the set of factors or genes (or genotype), *APP* a function that assigns to individuals their appearance or phenotype, *MAT* a function of crosses that assigns to any pair of parents its progeny, *DIST* the relative frequencies of the characteristics observed in the progeny, *DET* the relations postulated between genes and characteristics, and *COMB* the distributions of probability of genes in the progeny, which satisfy the law of matching. More formally expressed, the law establishes that if x = <*J, P, G, APP, MAT, DIST, DET, COMB*>, x is a problem-solving scheme, paradigm, shared example, exemplar or model of classical genetics if and only if for all $i, i' \in J$ such that MATOR is defined for $<i, i'>$ and for all $\gamma, \gamma' \in G$ such that $DET(\gamma) = APP(i)$ and $DET(\gamma') = APP(i')$: $COMB(\gamma, \gamma') = DIST(DET(\gamma), DET(\gamma'))$.

It is easy to see that in the law of matching we can identify the characteristics or "symptoms" of fundamental laws indicated in the previous section. First, the law of matching can be seen as a synoptic law because it establishes an substantial connection between the most important terms of genetics. It contains all the important terms that occurs in genetics both the genetics-theoretical ones (the set of the factors or genes, the distributions of probability of the genes in the progeny and the postulated relations between genes and characteristics) and the genetics-non-theoretical ones, which are empirically more accessible (individuals, the set of characteristics, the assignment of characteristics to individuals and of progeny to parental individuals, and the relative frequencies of characteristics observed in the progeny). Second, the law of matching is highly schematic and general and it possesses so little empirical content that it is irrefutable (i.e. it has a "quasi-vacuous" character). If we then look at the empirically determined relative frequency of the characteristics as well as the theoretically postulated distribution of genes and set out to test what this one the law claims – namely: that the coefficients in the distribution of characteristics and of genes in the progeny are (approximately) equal –, all this test really amounts to is an exercise with pencil and paper that does not involve any empirical work. Nevertheless, as we would expect it in the case of a fundamental law, despite being irrefutable it provides a conceptual frame in which all special laws can be formulated; that is, special laws with an increasingly degree of specificity and with an ever more limited domain of application until we reach terminal specializations, whose associated empirical claims can be seen as particular testable and, eventually, refutable hypotheses.

In addition, it is important to observe that this law has implicitly been accepted in every application of the theory because the community has used it as a general background assumption, providing a starting point for the analysis of different distributions of characteristics and serving as a guide for dealing with the plethora of empirical situations that geneticists face. So the primary role of the law of matching is to guide the process of specialization, determining the ways in which it must be specified to obtain special laws. According to this law, in order to explain the distributions of the parental characteristics in the progeny, the following parameters must be specified: (a) the number of pairs of genes involved (one or more), (b) the way in which the genes are related to the characteristics (complete or incomplete dominance, codominance or epistasis), and (c) how the parental genes are distributed in the progeny (with combinations of genes with the same probability or not). When these three types of specifications are made, terminal special laws are obtained, and it is the empirical claims that are associated with these laws that are capable of direct empirical test. In case that these laws "survive" a test, which means that the introduced specifications turn out to be the appropriate ones, one can say that the intended applications have become 'successful' and that the empirical systems become 'models' of the theory.

As long as Mendel's Laws impose additional constrictions on the law of matching, thereby adding information that is not already contained in its highly schematic formulation and restricting its area of application (as for example, on having considered only a pair of factors or having considered to be more than one, but the same probability for any possible combination of parental factors), they must be considered "special laws".

A first way of restricting the law of matching that characterizes a large class of models concerns the way in which the parental genes are distributed in the progeny (given by the function *COMB*), postulating that on having determined genotypes of the progeny, all the combinations of factors have the same probability. Thus, we obtain a specialization that could be considered to be a *general form* of Mendel's Laws as long as it includes both the first and the second of Mendel's Laws.[24]

If we specify the law even further in by assuming that the number of pairs of factors or genes involved equals one, we obtain the simple case of the Mendel's First Law (or 'Law of Segregation'), which is concerned with monohybridism.

Finally, if besides the way in which the parental genes are distributed in the progeny and the number of pairs of factors or genes involved, we also specify the

---

[24] In the early days of the "Mendelism", it turned out to be difficult to separate explicitly what we nowadays refer to as the "Law of Independent Assortment", or "Mendel's Second Law", and the "Law of Segregation", or "Mendel's First Law". Hugo de Vries was the first who spoke about the "Law of Segregation of the Hybrids" (loi of disjonction des hybrides' in French and "Spaltungsgesetz der Bastarde" in German) as discovered by Mendel. Nevertheless, he spoke about segregation of characteristics – "caractères" in French and "Merkmale" in German – and not of factors or genes. This is because at the time the distinction between characteristics and traits on the one hand, and factors or genes on the other hand was not completely clear (see de Vries, 1900a, b). Another of the so-called "rediscoverers", Carl Correns, constitutes an exception in this sense, since he explicitly postulates in 1900 an hereditary unit or Anlage (after the terminology of his teacher and correspondent of Mendel, Carl von Nägeli 1884) for every character at the individuum and states that these come always in pairs at the somatic cells uses the expression "Mendel's Rule" ("Mendels Regel" in German) to refer not only to the de Vries' "segregation law" but also to what has later been dubbed "Mendel's Second Law" (see Correns, 1900). The first one who used the term "independent assortment" was Thomas Hunt Morgan (Morgan, 1913). In 1919 Morgan for the first time explicitly spoke about two laws, the law of the segregation and the law of independent assortment of the genes and he attributed their discovery to Mendel, referring to them as "Mendel's First Law" and "Mendel's Second Law" respectively (see Morgan, 1919).

way in which genes are related to characteristics (by introducing the notions of complete dominance, incomplete dominance, recessivity, codominance or epistasis), the terminal specializations are reached. These terminal specifications explain the phenotypic proportions 3:1. Instances of this specialization correspond to what in Sinnott and Dunn (1925, 40–42, 45–50, 85) consider to be paradigmatic examples of Mendel's First Law.

Another specialization that can be found in the literature is the one which, assuming the equal probability of all the combinations of factors in the distribution of parental factors in the progeny, postulates that the number of pairs of factors or genes involved is two. By means of this specification one characterizes what one normally means by "Mendel's Second Law" (or "Law of Independent Assortment") as applied to dihybridism (Sinnott and Dunn, 1925, 67–70).

Furthermore, if one specifies the concrete way in which genes are related with characteristics, we again obtain terminal specializations. In this way we obtain for example the first type of cases in those specializations in which every pair of factors determines a different character, but one with complete dominance for both pairs of factors, which explains the phenotypic proportion 9:3:3:1. Instances of this specialization correspond to what in Sinnott and Dunn (1925, 63–67) refer to as paradigmatic examples of the so-called "Mendel's Second Law".

Other specializations can also be obtained following the line current of argument. For example, the specification in which every pair of factors determines a different character but with complete dominance for one pair of factors and partial dominance or codominance for the other one (it explains the inheritance of the comb of fowls). And we can deal with the case in which every pair of factors determines the same character with different manners of dominance and epistasis (that explains the phenotype proportions 9:7, 9:3:3:4, 12:3:1, 13:3, 15:1) along the same lines. This line of specialization characterizes the so-called 'factors interaction'.

In other specializations, the number of pairs of factors or the concrete way in which the factors are related to the characteristics will be different. For instance, the number of pairs of factors might be equal to three and any pair of factors might determine a different character with complete dominance for every component of the genotype (explaining the phenotypic proportion 27:9:9:9:3:3:3:1), or, with the number of pairs of factors equals to three, but with each of them having additives effects, determining the same character with a constant variation. The latter specialization captures the case of "multiple factors" for the quantitative inheritance.[25]

A further specialization tackling the case in which there are two pairs of factors involved is that of linkage, where not all combinations of factors are equally probable but some of them – those who are "united" or "linked" – appear more frequently than others. If factors are so strongly linked that they are always transmitted together if they come from the same parent, their linkage is called *complete*.

[25] An explicit formulation of these specializations can be found in Lorenzano (1995).

In such a case only the original paternal genotypes should be found. Often, however, due to the phenomenon of "recombination", the linkage is not complete (recombination is the phenomenon by which also new types of combinations called "recombinations" appear). These, however, do not occur as frequently as the paternal types. This "linkage" phenomenon explains the exceptions to the Mendel's Second Law.[26]

6. CONCLUDING REMARKS

Taking the first textbook of classical genetics that clearly possess all the features that Kuhn identifies in a science textbook as a starting point, this paper was devoted to a discussion of the problem of the existence of fundamental laws in biology. To begin with it was observed that problem-solving schemes, shared examples, exemplars, paradigms or models occupy a center stage. Then the hypothesis that the different problem-solving schemes, shared examples, exemplars, paradigms, or models of one and the same theory are tied together by structural similarity was explored and put on a firm foundation by using the structuralist concept of fundamental law. Based on this, the fundamental law of classical genetics, the law of matching, was identified. This satisfies all weak necessary conditions that a something needs to satisfy according to the structuralist view in order to be a law. This claim stands in stark contrast to the received wisdom according to which biology in general and genetics in particular do not have fundamental laws. A fortiori it is not true that (all) theories (if any) within the biological (and/or biomedical) sciences have a structure that is essentially different from the one of theories within the physical sciences. In fact, it has been shown that the structuralist view of fundamental laws and theories – initially coined within the context of a philosophical analysis of physical theories – also applies to the domain of genetics, which shows how fruitful this approach is. It is also worth pointing out that this case study suggests that the possibility of identifying fundamental laws in biology need not to be limited to classical genetics, as has it has been suggested in the context of the debates over the status of the Principle of Natural Selection, in particular by authors such as Brandon (1978, 1981). On the other hand, the Hardy-Weinberg law, discussed by Sober (1984, 1993, 1997) and Elgin (2003), can hardly be considered the fundamental law of population genetics. Instead, it can be considered as a special law; rather, its relation to the fundamental law it could be comparable to the relation that the principle of inertia bears to Newton's Second Principle. But unlike Newtonian mechanics, population genetics still waits an explicit statement of its fundamental law that occupies the same systematic position in the theory as Second Principle in classical particle mechanics. However, only a detailed analysis of these and other biological theories will be able to decide whether they are cases analogous to the one discussed in this paper.

---

[26] For details see Lorenzano (1995).

ACKNOWLEDGEMENTS

This work has been partially founded by the research projects PICTR2002 N° 00219 (ANPCyT) and PICT2003 N° 14261 (ANPCyT). I would like to thank Ricardo Gómez (California State University) for the comments he made at a version of this paper read at the APA Central Division Meeting held in Chicago, in April 2005. I would also like to thank Roman Frigg (London School of Economics) for his helpful suggestions and for improving my English.

REFERENCES

Balzer W (1979a) Die epistemologische Rolle des zweiten Newtonschen Axioms. Philosophia Naturalis 17:131–149.

Balzer W (1979b) Logische versus physikalische Definitionen in der physikalischen Begriffsbildung In: Balzer W, Kamlah A (eds) Aspekte der physikalischen Begriffsbildung. Theoretischer Begriffe und operationale Definitionen, Friedr. Vieweg and Sohn, Braunschweig/Wiesbaden, 13–36.

Balzer W, Dawe CM (1990) Models for genetics. München Institut für Philosophie, Logik und Wissenschaftstheorie.

Balzer W, Lorenzano P (1997) The logical structure of classical genetics. Zeitschrift für allgemeine Wissenschaftstheorie 31:243–266.

Balzer W, Sneed J (1977/1978), Generalized Net Structures of Empirical Theories I and II. Studia Logica 36:195–211; 37, 167–194.

Balzer W, Moulines CU, Sneed J (1987) An architectonic for science. The Structuralist Program, Reidel, Dordrecht.

Balzer W, Moulines CU, Sneed J (eds) (2000) Structuralist knowledge representation: paradigmatic examples, Rodopi, Amsterdam.

Bartelborth Th (1988) Eine logische Rekonstruktion der klassischen Elektrodynamik. Peter Lang, Frankfurt am Main.

Beatty J (1980) Optimal-design models and the strategy of model building in evolutionary biology. Philos Sci 47:532–561.

Beatty J (1981) What's Wrong with the Received View of Evolutionary Theory? In: Asquith PD, Nickles T (eds), PSA 1980, Philosophy of Science Association, vol. 2, East Lansing, Michigan, 397–426.

Beatty J (1995) The evolutionary contingency thesis. In: Wolters G, Lennox J (eds) Theories and rationality in the biological sciences, the second annual pittsburgh/konstanz coloquim in the philosophy of science, University of Pittsburgh Press, Pittsburgh, PA, pp 45–81.

Bennett AJ (1964) Mendel's Laws? School Science Review 46:35–42.

Bennett AJ (1965) Mendel and Mendelism: A Possible Reinterpretation. Journal of the Institute of Biology 12:72–74.

Beurton PJ, Falk R, Rheinberger H-J (eds) (2000) The concept of the gene in development and evolution, historical and epistemological perspectives. Cambridge University Press, Cambridge.

Bowler P (1989) The mendelian revolution. The Athlone Press, London.

Brandon RN (1997) Does Biology Have Laws? The Experimental Evidence, Philos Sci 64, Proceedings, S444–S457.

Brandon RN (1978) Adaptation and evolutionary theory. Stud Hist Phil Sci 9:181–206.

Brandon RN (1981) A structural description of evolutionary biology. In: Asquith PD, Nickles T (eds) PSA 1980, Philosophy of Science Association, vol 2, East Lansing, Michigan pp 427–439.

Cadevall i Soler M (1988) La estructura de la teoría de la evolución. Universitat Autònoma de Barcelona, Bellaterra.

Correns C (1900) Gregor Mendels Regel über das Verhalten der Nachkommenschaft der Bastarde. Berichte der Deutschen Botanischen Gesellschaft 18:158–168.

Darden L (1991) Theory change in science. Strategies from Mendelian genetics. Oxford University Press, Oxford.
Darden L (1996) Essay review. Generalizations in biology. Kenneth F Schaffner, Discovery and explanations in biology and medicine, University of Chicago Press, Stud Hist Phil Sci 27(3):409–419.
Dawe CM (1982) The structure of genetics, Doctoral Dissertation, University of London, London.
Díez JA, Lorenzano P (2002) La concepción estructuralista en el contexto de la filosofía de la ciencia del siglo XX. In: Díez JA, Lorenzano P (eds) Desarrollos actuales de la metateoría estructuralista: problemas y discusiones, Quilmes: Universidad Nacional de Quilmes/Universidad Autónoma de Zacatecas /Universidad Rovira i Virgili, pp 13–78.
Elgin M (2003) Biology and a priori laws. Philos Sci 70:1380–1389.
Ereshefsky M (1991) The semantic approach to evolutionary theory. Biology and Philosophy 6:59–80.
Fleck L (1935) Entstehung und Entwicklung einer wissenschaftlichen Tatsache. Einfuhrung in die Lehre vom Denskstil und Denkkollektiv, Benno Schwabe and Co, Basel.
Fodor J (1974) Special sciences (or: the disunity of science as a working hypothesis). Synthese 28:97–116
Fodor J (1991) Hedged laws and psychological explanations. Mind 100:19–33.
Gould SJ (1989) Wonderful life. The burgess shale and the ature of history, WW. Norton and Company, New York.
Gould SJ (2002) The structure of evolutionary theory. The Belknap Press of Harvard University Press, Cambridge, MA.
Hempel CG, Oppenheim P (1948) Studies in the logic of explanation. Philos Sci 15:135–175.
Hull D (1974) Philosophy of biological science. Prentice-Hall, Englewood Cliffs, NJ.
Keller EF (2000) The century of the gene. Harvard University Press, Cambridge, MA.
Kitcher P (1984) (1953) And all that: a tale of two sciences. The Philos Rev 93:335–373.
Kitcher P (1989) Explanatory unification and the causal structure of the world. In: Kitcher P, Salmon WC (eds) Scientific explanation, minnesota studies in the philosophy of science, vol 13. University of Minnesota Press, Minneapolis, pp 410–505.
Kyburg HE (1968) Philosophy of science: a formal approach. The Macmillan Company, New York.
Kuhn TS (1959) The essential tension: tradition and innovation in scientific research. In: Taylor CW (ed) The third (1959) university of utah research conference on the identification of scientific talent, University of Utah Press, Salt Lake City, pp 162–174; reissued in The essential tension. selected studies in scientific tradition and change, University of Chicago Press, Chicago, 1977, pp 225–239.
Kuhn TS (1962/1970) The structure of scientific revolutions. University of Chicago Press, Chicago.
Lewontin RC (1974) The genetic basis of evolutionary change. Columbia University Press, New York.
Lindenmayer A, Simon N (1980) The formal structure of genetics and the reduction problem. In: Asquith PD, Giere RN (eds) PSA 1980, Philosophy of Science Association, East Lansing, Michigan, pp 160–170.
Lloyd E (1988) The structure and confirmation of evolutionary theory. Greenwood Press, New York.
Lorenzano P (1995) Geschichte und Struktur der klassischen Genetik. Peter Lang, Frankfurt am Main.
Lorenzano P (1997) Hacia una nueva interpretación de la obra de Mendel. In: Ahumada J, Morey P (eds) Selección de trabajos de las VII Jornadas de Epistemología e Historia de la Ciencia, Córdoba: Facultad de Filosofía y Humanidades, Universidad Nacional de Córdoba, pp 220–231.
Lorenzano P (2000) Classical genetics and the theory-net of genetics. In: Balzer W, Moulines CU, Sneed J (eds) Structuralist knowledge representation: paradigmatic examples, Rodopi, Amsterdam, pp 251–284.
Lorenzano P (2001) On biological laws and the laws of biological sciences. Rev Pat Fil 2:29–43.
Lorenzano P (2002a) La teoría del gen y la red teórica de la genética. In: Díez JA, Lorenzano P (eds) Desarrollos actuales de la metateoría estructuralista: problemas y discusiones, Quilmes: Universidad Nacional de Quilmes/Universidad Autónoma de Zacatecas /Universidad Rovira i Virgili, pp 285–330.
Lorenzano P (2002b) Leyes fundamentales, refinamientos y especializaciones: del 'Mendelismo' a la 'teoría del gen'. In: Lorenzano P, Molina FT (eds) Filosofía e Historia de la Ciencia en el Cono Sur, Universidad Nacional de Quilmes, Quilmes, pp 379–396.

Lorenzano P (2002c) Kenneth Schaffner, las teorías de alcance intermedio y la concepción estructuralista de las teorías, communication presented at the conference III Encuentro Iberoamericano sobre Metateoría Estructural, Granada, España.
Lorenzano P (2006) Fundamental Laws and Laws of Biology. In: Ernst G, Niebergall K-G (eds.), Philosophie der Wissenschaft – Wissenschaft der Philosophie. Festschrift für C.Ulises Moulines zum 60. Geburstag, Mentis-Verlag, Paderborn, 129–155.
Lorenzano P (2006) La emergencia de un programa de investigación en genética. In: Lorenzano, P, Martins, LA-CP, Regner, AC (eds) Ciências da vida: estudos filosóficos e históricos, Associação de Filosofia e História da Ciência do Cone Sul (AFHIC), Campinas, pp 333–360.
Mayr E (1991a) Eine neue Philosophie der Biologie. Wissenschaftliche Buchgesellschaft, Darmstadt.
Mayr E (1991b) One long argument: charles darwin and the genesis of modern evolutionary thought. Harvard University Press, Cambridge.
Morange M (2001) The misunderstood gene. Harvard University Press Cambridge, MA.
Morgan TH (1913) Heredity and sex. Columbia University Press, New York.
Morgan TH (1919) The physical basis of heredity. Lippincott, Philadelphia.
Moya A (1989) Sobre la estructura de la teoría de la evolución. Anthropos, Barcelona.
Mosterín J (1984) Conceptos y teorías en la ciencia. Alianza, Madrid, 2nd edn. 1987
Moulines CU (1978/1982) Forma y función de los principios-guía en las teorías físicas. In: Exploraciones metacientíficas, Alianza, Madrid, pp 88–107 (first publication as Cuantificadores existenciales y principios-guía en las teorías físicas. Crítica 10 (1978):59–88. Translated as Existential quantifiers and guiding principles in physical theories. In: Gracia JJE, Rabossi E, Villanueva E, Dascal M (eds) Philosophical analysis in latin america, Reidel, Dordrecht, 1984:173–198).
Moulines CU (1991) Pluralidad y recursión. Alianza, Madrid.
Munson R (1975) Is biology a provincial science? Philos Sci 42:428–447.
Nägeli Cv (1884) Mechanisch-physiologische Theorie der Abstammungslehre. R. Oldenburg, München and Leipzig.
Olby R (1979) Mendel no Mendelian? History of Science 17:53–72; reissued in Olby R (1985) Origins of Mendelism. The University of Chicago Press, Chicago, 2nd edn (extended), pp 234–258.
Olby R (1979) Mendel no Mendelian? History of Science 17:57–72. Reissued in: Olby R (1985) Origins of Mendelism, The University of Chicago Press, Chicago, 2nd edition.
Rizzotti M, Zanardo A (1986) Axiomatization of genetics I and II. J Theor Biol 118:61–71, 145–172.
Rosenberg A (1985) The structure of biological science. Cambridge University Press, Cambridge.
Ruse M (1970) Are there laws in biology? Australas J Philos 48:234–246.
Ruse M (1973) The philosophy of biology. Hutchinson, London.
Salmon WC (1989) Four decades of scientific explanation. In: Kitcher P, Salmon WC (eds) Scientific explanation, minnesota studies in the philosophy of science, vol 13, University of Minnesota Press, Minneapolis, pp 3–219.
Sarkar S (1998) Genetics and reductionism. Cambridge University Press, Cambridge.
Schaffner KF (1980) Theory structures in the biomedical sciences. The J Med Philos 5:57–97.
Schaffner KF (1986) Exemplar reasoning about biological models and diseases: a relation between the philosophy of medicine and philosophy of science. The J Med Philos 11:63–80.
Schaffner KF (1993) Discovery and explanations in biology and medicine. University of Chicago Press, Chicago and London.
Schlick M (1918) Allgemeine erkenntnislehre. Julius Springer, Berlin
Sinnott EW, Dunn LC (1925) Principles of genetics: an elementary text, with problems. McGraw-Hill, New York, 2nd edn. 1932; 3rd edn. 1939; with T Dobzhansky as co-author, 4th edn. 1950; 5th edn. 1958.
Sintonen M (1991) How evolutionary theory faces the reality. Synthese 89:163–183.
Smart JJC (1963) Philosophy and scientific realism. Routledge and Kegan Paul, London.
Sneed JD (1971) The logical structure of mathematical physics. Reidel, Dordrecht, 2nd revised edn. 1979.
Sober E (1984) The nature of selection: evolutionary theory in philosophical focus. MIT Press, Cambridge, MA.

Sober E (1993) Philosophy of biology. Westview Press, Boulder.
Sober E (1997) Two outbreaks of lawlessness in recent philosophy of biology. Philos Sci 64, Proceedings, S458–S467.
Stegmüller W (1973) Theorienstrukturen und Theoriendynamik. Springer, Heidelberg.
Stegmüller W (1976) Eine 'subjektivistische' Variante des Begriffs der physikalischen Theorie. In: Neue Wege der Wissenschaftsphilosophie, Springer, Berlin/Heidelberg/New York 1980, pp 56–86.
Stegmüller W (1978) A combined approach to the dynamics of theories. Boston studies in the philosophy of science 59/136:151–186.
Stegmüller W (1979a) The structuralist view: survey, recent development and answers to some criticisms. In: Niiniluoto I, Tuomela R (eds) The logic and epistemology of scientific change (Acta Philosophica Fennica 30), North-Holland, Amsterdam, pp 113–129.
Stegmüller W (1979b) The structuralist view of theories. Springer, Berlin.
Stegmüller W (1983) Erklärung–Begründung–Kausalität, Springer, Berlin/Heidelberg/New York, 2nd edn. extended and modified.
Stegmüller W (1986) Theorie und Erfahrung, Band II, Dritter Halbband, Springer, Berlin/ Heidelberg/New York.
Swartz N (1995) The neo-humean perspective: laws as regularities. In: Weinert F (ed) Laws of nature. essays on the philosophical, scientific and historical dimensions, de Gruyter, Berlin, pp 67–91.
Thompson P (1989) The structure of biological theories. State University of New York Press, Albany, NY.
Tuomi J (1981) Structure and dynamics of darwinian evolutionary theory. Syst Zool 30:22–31.
Tuomi JyEH (1979) An analysis of natural selection in models of life-history theory. Savonia 3:9–16.
van Fraassen B (1977) The only necessity is verbal necessity. J Philos 74:71–85.
van Fraassen B (1989), Laws and symmetry. Clarendon Press/Oxford University Press, Oxford.
van Fraassen B (1993) Armstrong, Cartwright, and Earman on *Laws and Symmetry*. Philos Phenomenol Res LIII (2):431–444.
Vries H de (1900a) Sur la loi de disjonction des hybrides. Comptes Rendus de l'Académie des Sciences 130:845–847.
Vries de H (1900b) Das Spaltungsgesetz der Bastarde. Ber Deut Botan Ges 18:83–90.
Wassermann GD (1981) On the nature of the theory of evolution. Philos Sci 48:416–437.
Weinert F (ed) (1995) Laws of nature. Essays on the philosophical, scientific and historical dimensions, de Gruyter, Berlin.
Williams MB (1970) Deducing the consequences of evolution: a mathematical model. J Theor Biol 29:343–385.
Wittgenstein L (1953) Philosophische untersuchungen/philosophical investigations, Basil Blackwell, Oxford.
Woodger JH (1959) Studies in the foundations of genetics. In: Henkin L, Suppes P, Tarski A (eds) The axiomatic method, North-Holland, Amsterdam, pp 408–428.

CHAPTER 8

EPI-GENETICIZATION

Where biological and philosophical thinking meet

LINDA VAN SPEYBROECK*, GERTRUDIS VAN DE VIJVER†
AND DANI DE WAELE

Ghent University, Department of Philosophy and Moral Science Centre for Critical Philosophy, Ghent, Belgium

**linda.vanspeybroeck@UGent.be*

†gertrudis.vandevijver@urgent.be

Abstract: This volume discusses how contemporary genetics is present in and has an influence on domains other than the strict genetical or biological field of knowledge. It thus focuses on 'the geneticization' of (scientific) thinking. In order to give meaning to this concept, it needs to be addressed what contemporary genetics stands for and wherein its conceptual and practical approach towards living systems lies. Philosophy of biology has a long history in taking classical and molecular genetics under analysis, and has often described genetics as gene-centric and reductionist. Here, it is argued that this analysis has become outdated and that today 'geneticization' should be interpreted as 'epi-geneticization'. This conceptual shift is supported by experimental research in molecular biology itself, showing that molecular biology is already taking up the challenge of approaching biology in less gene-centric terms. This holds implications for philosophy of biology in its debate on the gene concept, in particular, and in its study of scientific perspectives on biological organization, in general

1. GENETICIZATION AS IN "IT'S ALL IN THE GENES"

The risk of speaking about a 'geneticization' of human thinking without further ado, lies in the presupposition that current genetics as an experimental and conceptual approach towards living organisms is to be equated with *gene-centrism*, *genetic reductionism* or even with *genetic determinism*. In general, the *gene-centric* version of molecular biology puts genes central in the explanation and understanding of biological processes and organization, up to the point that the role of other factors is highly neglected or is considered as less important. Instead, genes are set apart as crucial factors attributed with essential characteristics such as containing phenotypic information or being the heritable units of life. Gene-centrism is an epistemic

A. Fagot-Largeault et al. (eds.), The Influence of Genetics on Contemporary Thinking, 115–133.

concept – or a knowledge perspective – referring to the way humans think about biological organization and model it accordingly. More specifically, *genetic reductionism*[1] stresses the molecular analysis of the genetic level as the best methodology to approach a living organism. This level is considered the best causal starting point for a study of biological processes, as the essential characteristics of an organism are thought to reduce hereto. A stronger version hereof is *genetic determinism*, taking an explicit focus on the genetic level as necessary and sufficient because genes are ontologically seen to contain all information to build the entire organism. Genes determine the organism, i.e. the complexity of higher levels in an organism comes forth from interactions at the genetic level. Historically, the idea that *life is preformed in the genes* can be situated within the preformationist tradition – a dominant line of thought presupposing the existence of units that either *in actuality* or *potentially* preform organisms in all their characteristics (Van Speybroeck et al., 2002). This presupposition also lies at the basis of the search for the material characterization of these units. In this search, the term *gene* changed from referring to cellular units in embryology over abstract units in Mendelian genetics to concrete DNA-sequences in molecular biology (Morange, 1998).[2]

The combined use of a *DNA-oriented gene-concept* and a *gene-centric perspective* gained huge popularity through the work of Richard Dawkins. While Dawkins interpreted a gene as a sequence of nucleotides "lying between a START and an END symbol, and coding for one protein chain" (Dawkins, 1989, 28) – long enough to be functional, short enough to survive recombination throughout the next generations – he also interpreted it as the ultimate unit of development, natural selection and evolution (Beurton, 2000). For example, in claiming that "the gene determines a protein sequence that influences X that influences Y that influences Z that eventually influences the wrinkliness of the seed or the cellular wiring up of the nervous system" (Dawkins, 1989, 240), he presupposed that genes fully determine development and are in fact units of organismal identity. Hence his interpretation of development in terms of a calendar or a clock (cf. Dawkins, 1989, 262), pleading for a determined and rigorous account in which genes are *first movers*, the *prior causal and informative factors* unleashing a predetermined linear cascade. As in this view only DNA mutations are considered important enough in causing an alteration to this causal chain, developmentalist accounts on organisms tend to be black boxed. Organisms are merely "survival machines – robot vehicles blindly programmed to preserve the selfish molecules known as genes" (Dawkins, 1989, v).

Dawkins's view can be seen as a reiteration of molecular biology's *Central Dogma*, which presents biological hereditary information[3] as spreading via a

---

[1] For comments on how the philosophical treatment of reductionism has been of use to biology, see Robinson, 1992, 466–470 and Sarkar, 1998.

[2] For an account of how orthodox philosophy of biology describes the gene concept in classical genetics and in molecular biology, see Waters (1994), who himself gives an interesting alternative view.

[3] Information here stands for "the precise determination of sequence, either of bases in the nucleic acid or of amino-acid residues in the protein" (Crick, 1958, 153).

one-way direction. That is, from DNA to RNA, and from RNA to protein, not *vice versa* (Sarkar, 1996). Throughout the years, this scheme not only structured "our basic representations and models of what counts as genetic" (Griesemer, 2002, 98), it also focused biological thinking and acted to eliminate or suppress alternative habits of thought (Morange, 1998). For example, while promoting a reductionist molecular approach, the Dogma provided a shift in focus from protein to DNA. Indeed, whereas Crick's original stress on "the central biochemical importance of proteins" (Crick, 1958, 152) was kept down, the dominating role of genes "and the genetic linearity within the functional gene" (Crick, 1958, 152) moved centre-stage. Also, the linear causality from DNA to protein found its place in the context of natural selection, random genetic mutations and the impossibility to cross the Weismannian Barrier between soma and germ. This created a climate almost *a priori* disinterested in ideas on environments changing proteins and proteins influencing DNA (Gilbert, 1996).

Pushed to its logical conclusions, the Dogma symbolizes the idea that "to decode the [genetic] information is to know the organism" (Morange, 1998, 76) – which became the incentive of the many Genome Projects in the 1980s. This idea is the corner stone of a strong molecular gene-centrism: (i) in terms of DNA, genes are portrayed as the essential heritable molecules in biological organization, (ii) the organism no longer counts as a biological unity, (iii) the environment is restricted to selective pressures and random insults at the DNA level. As such, the Dogma not only became the gene-centric touchstone of *molecular* biology (Torres, 1999), but of biology in general.

2. GENETICIZATION AS "IT'S NOT ALL IN OUR GENES"

During the past decade, philosophy of biology has been criticising genetics and has argued to trade the gene-centric stance 'it's all in our genes' for 'it's not all in our genes' (cf. Moss, 2003) or even for 'it's not in our genes at all' (cf. Oyama et al., 2001). Today, both the current status of the Central Dogma[4] and Dawkins's notion of *selfish genes* continue to be attractive poles for re-evaluation. In essence, admitting (i) that not the genes themselves but their effects (i.e. the vehicles or the organismal soma) are under direct and immediate influence of natural selection (Dawkins, 1989, 235), (ii) that "all that genes can really influence directly is protein synthesis" (Dawkins, 1989, 240), and (iii) that gene activity in embryonic development depends on a specific spatio-temporal context, demonstrates that also

[4] When asking whether today it is still justified to consider Crick's *dictum* a true *Dogma*, it is instructive to read the original publication of Crick in 1958. The Dogma is presented as an "oversimplified" scheme for the general reader (Crick, 1958, 154), which intended merely to be a heuristic tool for the development of new hypotheses on gene expression. This allowed speculations on RNA–RNA and RNA–DNA interactions and on translations of proteins directly from DNA, and did neither diminish the role of proteins, nor extrapolate the 'control' of DNA to all biological levels, nor isolate DNA function from its cellular context (cf. Thieffry, 1998).

Dawkins's gene-centrism stands in tension with the upcoming image of the organism as an individual, coherent, integrated and complex whole.

This tension, which also plays a prominent role in social debates on biological topics such as genetically modified organisms, medical care and cloning (Nelkin and Lindee, 1995), has been extensively discussed, giving room to a renewed attention for the *de novo* epigenesis or development of organisms and for the complex of internal and external causal nodes impinging on it (cf. Lewontin, 2001; Keller, 1995; Ruiz-Mirazo et al., 2000). In this debate, philosophers have made ample use of an ever-growing amount of experimental data and conceptual classifications provided by molecular biology. To name but a few historical examples in this regard, *DNA editing*, discovered in the mitochondria of unicellular trypanosomes, allows non-functional RNAs to be 'freshened up' after transcription from DNA through the addition of the missing nucleotides by repair mechanisms. Building on research in the 1950s on the tobacco mosaic virus, the existence of an information flow from RNA to DNA (or *reverse transcription*) was confirmed, directly assaulting the Central Dogma and making DNA less inviolable. A similar conclusion was drawn in the study of *pseudogenes* (which resemble protein-coding genes, but are non-active and without introns, and probably result from a reverse transcription from RNA to DNA), supporting the idea that the genome could integrate information from the cytoplasm, "raising with it the specter of Lamarckism" (Morange, 1998, 211). Also the change in the 1970s from unicellular prokaryotic to multicellular eukaryotic model organisms undermined the image of DNA as static and solid. By the late 80s, through the discovery of *differential RNA splicing*, *gene overlaps* and diverse *mRNA editing mechanisms*, it became clear that it is often problematic to predict which protein will be translated from a given DNA sequence. Speculations about how *prions* could induce the expression of a silent aberrant host gene or catalyse post-translational changes in normal proteins further opened the discussion about adding a protein-DNA or a protein-protein link to the Dogma (Thieffry, 1998).

These examples not only illustrate that if DNA still is to *determine* a protein sequence, it does so in a non-linear manner which no longer can be seen as separate from regulatory proteins and other non-DNA factors or from processes playing at higher organizational levels (cf. Waters, 1994). It also shows that molecular biology *itself* implicitly or explicitly has grown to challenge gene-centrism. Specifically, although historically grown out of the quest to materially *isolate* genetic units, this 'new molecular biology' allows questioning if genes can be *atomized* or singled out as unmoved movers in a linear causal chain. This puts gene-centrism at stake because as soon as genes are denied their separate physical or functional status inside the organism, the *system* in which they reside is to be acknowledged.

In this regard, philosophers of biology have acknowledged the importance of the complexity paradigm in which the idea of *circular casuality* is seen as a useful alternative or addendum to the performationist idea of *linear casuality* (cf. Van de Vijver et al., 2003 for a review). Together with a focus on *process* and *interaction*, thinking in terms of complexity contributes to examine a de-centralization of the gene-concept. However, evading preformationism also means taking into account 'holistic' obstacles, such as *can one still delimit a (developing) system?* For example, are environmental factors external to the 'traditional' organismal boundaries part of the system or not? Also, as soon as causal determination is interpreted as complex and diversified with forms of so-called *downward causation* in which the whole constrains the lower level parts in their (inter)actions, the question arises whether the most adequate study of a living system will be a study in real-time while the developmental processes are taking place. Studies in complexity minimally suggest that the system's 'potential' is out of reach of certain types of description and explanation, and that not all of its future states and behaviour are predictable. Therefore, the relevancy and the limits of current methodologies used in analysing living systems is under discussion. Does this mean however that only the 'actual' of the organism, followed in the real-time development, should come in focus? Next, 'de-centralizing the gene' does not mean 'forgetting about genes'. That is, besides clarifying how the system takes up former so-called pure genetic functions, one also has to account for the presence and functioning of genes in this larger system.

In this regard, *epigenetics* forms an interesting case study of the range a biological discipline can cover as an alternative to gene-centrism.

3. EPI-GENETICIZATION AS "GOING BEYOND" GENE-CENTRISM

3.1. Epigenetics *versus* Genetics

The neologism *epigenetics*, introduced in the 1940s by the embryologist Conrad Hal Waddington (1947/1940, 1975), presents itself as a study *going beyond* genetics. However, exactly *what* is 'epi' about epigenetics (cf. Griesemer, 2002, 97)?

When Wu and Morris (2001) report that neither of them "has ever been quite sure of the meaning" of this word, this is because epigenetics is hard to grasp when not considering the strength of the sieve *genetics* uses or has used to withdraw certain phenomena from being studied. It is instructive to recall that William Bateson coined genetics in 1905 as a general name for the study of heredity and variation – a study originally covering plenty of model organisms from flowering plants, poultry, rabbits, sea urchins, beetles and snails, to starfish. Later on, genetics was diminished to study a few model organisms easy to maintain under non-insulting laboratory conditions and with genotypes generally considered to obey Mendelian laws (Bolker, 1995). Because of this choice, genetics – strengthened by the impact of the Central Dogma as a general framework – left us with a biased view, *automatically* opening up the perspective of an *epi*-genetics to account for those findings not fitting the genetic scheme. In this sense, epigenetics pre-exists Waddington's neologism simply by the fact that genetics does not present (nor intends to present) a complete theory of biological organization. Nowadays, as argued above, old-school genetic reductionism in the sense of "complex properties of the organism are explained by those of one or a few molecular components" becomes more and more inappropriate, even within genetics (Morange, 2002, 57). As such, the gap between *genetics* (as the study of the transmission and processing of heritable elements) and *epigenetics* seems to diminish.

Nonetheless, the distinction between both terms is kept intact in the academic literature. According to Morange (2002), this has to do with the fact that what genetics fully incorporated in its discourse and practice, no longer tends to be addressed by epigenetics. However, that genetics is historically casted on a gene-centric mould also plays. Due to its more restricted conceptual load, orthodox genetics is much slower – or more stubborn – to incorporate 'deviating' insights and to develop more modern connotations. This also explains why plenty of the epigenetic phenomena discovered and described as early as the 1930s, today remain to stand outside of genetics as exceptions and play no role in its theory.[5] Also, today the urge to denote research as *epi*-genetic – and thus as *exceptional* to the classical genetic perspective – is bigger than in the 1930s, when phenomena such as structural inheritance of the cortex in unicellular organisms were considered without much ado as an extranuclear form of heredity (Morange, 2002). This confirms that *genetics* is still often conceived of in terms of the narrow image on heredity as DNA-centred and *development* as instructed by the genetic program.[6] In this regard, epigenetics can be seen to constantly challenge genetics 'to keep up', creating a space in which concepts evolve flexibly according to changes in the scientific landscape.

In sum, only when genetics is interpreted as a bastion of gene-centrism, epigenetics fully comes of meaning. As such, biological discourse finds in molecular epigenetics the opportunity to lay out the fundaments of a conceptual framework on biological organization alternative to gene-centrism, helping to promote the shift from a 'geneticization' to an 'epi-geneticization' of thinking.

3.2. Evolving Definitions

Epigenetics is complexified because of its own conceptual evolution, throughout which it changed from an abstract reference to embryological epigenesis[7] to a concrete molecular study of gene regulation and inheritance. This is demonstrated in the diverse definitions of epigenetics. These definitions also show how epigenetics purports to evade gene-centrism.

In the 1940s, Waddington differentiated epigenetics from genetics by theoretically labelling the former the study of all causal processes and mechanisms impinging on

---

[5] F. Ex. morphological heredity depending on the cellular cortex (and not on the nucleus or on DNA present in cytoplasmic organites) as discovered by Janine Beisson and Tracy Sonneborn in unicellular organisms such as *Paramecium aurelia* finds no place in our genetic theories because it cannot be explained directly by gene action (Jablonka and Lamb, 1995).

[6] Also philosophers and historians of biology, in their focus on theory-based approaches towards biology, have contributed to this narrow view on genetics (Waters, 2004).

[7] "The mysterious workings of Nature that allow structure to form *de novo* from the apparent structureless mass that results from the union of egg and sperm" (Wu and Morris, 2001). See also Van Speybroeck et al., 2002.

the genotype during the epigenesis of the phenotype. Epigenetics thus stood for the whole of development itself and involved a holistic view on biological organization (Van Speybroeck, 2002). Today, this stance is reused in the definition of epigenetics as "*everything that leads to the phenotypic expression of the genetic information in an individual*" (Jablonka and Lamb, 1995, 80). However, by using concepts such as *genetic information* and *phenotypic expression*, this definition evokes the traditional genotype-phenotype distinction. In this regard, the definition on which currently consensus is reached is more careful, describing epigenetics as *"the study of mitotically and/or meiotically heritable changes in gene function that cannot be explained by changes in DNA sequence"* (Riggs et al., 1996, 1). Often it is added that these changes are *important for the understanding of the developmental processes and phenotypic traits of the organism* (cf. Henikoff and Matzke, 1997; Lewin, 1998).[8]

This consensus definition is much indebted to the molecular biologist Robin Holliday (Wu and Morris, 2001). In 1987, in reference to Waddington, he defined genetics as the study of the transmission of genes, and epigenetics as the study of changes in gene activity during development. Stress was on epigenetic switches turning particular genes on or off during developmental processes, producing transient changes in gene activity or permanent patterns of activities (Holliday, 1987). In 1990, Holliday continued to describe epigenetics as the study of the mechanisms that impart temporal and spatial control on the activities of genes required for the development of a complex organism from the zygote to the adult. Nonetheless, he began to stress the notion of *epigenetic inheritance* by claiming that "the mechanisms of epigenetic control must include the inheritance of a particular spectrum of gene activities in each specialized cell" (Holliday, 1990 in Jablonka and Lamb, 2002, 87). In addition to the classical DNA code, he also found it necessary "to envisage the superimposition of an additional layer of information which comprises part of the hereditary material" (ibid., 87). By 1994, the "unfolding of the genetic programme" was described as "ultimately depend[ing] on the activation or inactivation of specific genes, or the interactions between genes and the products of genes" (Holliday, 1994, 453). Epigenetics was redefined as the study of "changes in gene expression, which occur in organisms with differentiated cells, and the mitotic inheritance of given patterns of gene expression [i.e. stable mitotic inheritance of given patterns of gene activity is a key feature of epigenetic control]" (ibid., 453). This new definition also applied to adult organisms existing out of already differentiated cells. Hereby, Waddington's *embryological* conception of epigenetics was extended to a broader view on development, including physiological genetics. Because this definition did not contain any restriction on the nature of the mechanisms involved, also changes in DNA sequence (such as

[8] A related view on epigenetics: the study of "modifications in gene expression [either transcriptionally or posttranscriptionally] that are controlled by [mitotically and/or meiotically] heritable but potentially reversible changes in DNA methylation and/or chromatin structure" (Petronis, 2001, 143).

classical mutations and DNA rearrangements) appeared included as mechanisms for changing gene expression. However, in the same article, while re-establishing an explicit link between epigenetics and meiotic inheritance by stressing that a "supplementary definition of epigenetics [is] to include transmission of information from one generation to the next" (Holliday 1994, 454), Holliday restricted epigenetics to (i) heritable phenomena "other than the DNA sequence itself" (ibid.) and (ii) "nuclear inheritance which is not based on differences in DNA sequence" (ibid.). In doing so, he explicitly placed epigenetics at a distance from orthodox genetics. This stress is kept in the consensus definition, which comprises a fusion of Holliday's 1994 descriptions. Moreover, it no longer relies on a genotype-phenotype dichotomy, but incorporates two elements countering gene-centrism.

Firstly, although the establishment of a developmental theory seems to have become less central than in Waddington's era, the epigenetic stress on the regulation of gene function by non-DNA elements necessarily relies on a developmentalist framework. That is, while focusing on gene function, epigenetics is obliged to take an active interest in the diverse contexts and systems that impinge on the DNA-level, such as "processes of establishment, regulation, maintenance, and propagation of differentiated cellular states of *cells* in multicellular bodies. That is to say, the problem of epigenetics concerns the ways the differentiating and differentiated states of cells are established in development" (Griesemer, 2002, 102). Here, the classical genetic framework in which only the transmission of (changes in) DNA and a rough genotype-phenotype link stands central no longer suffices.

Secondly, while neglected in the genetic theory of heredity, the consensus definition explicitly focuses on *epigenetic inheritance across generations* in order to acknowledge that not all heritable information leading to the phenotype in this or next generations is directly 'inscribed' in nucleotide sequences (Martienssen and Colot, 2001). Not only DNA sequences undergoing permanent changes can be inherited, also less stable and even reversible changes in gene activity through non-DNA elements often induced by external influences can be heritable and may play a role in biological organization and evolution. In this respect, epigenetics revolts to the Central Dogma and complements the neo-Darwinian stress on the linear nucleotide sequence and on random DNA mutations by providing room for *interactivism.* This is done by portraying the realm of DNA as being much more interconnected with the realm of non-DNA than the genetic paradigm previously led to believe.

3.3. A Layered View on Genomic Contexts

Although epigenetics *per definition* reacts against orthodox genetics, it also builds upon it by placing *nuclear* gene function at its core. Epigenetics thus can be interpreted equally as being partially in opposition or as being partially 'entrenched' with genetics (Griesemer, 2002). Even more, while today a firm link is established between epigenetics and molecular biology – the science Waddington saw as most promising to study epigenetics –, this molecularization somehow stands in tension with the original view on epigenetics as a holistic concept. In contrast with Waddington's focus on gene networks residing in developing organisms, epigenetic research often seems to study the isolated regulation of one specific (trans)gene. This apparently resembles a new form of gene-centrism in which maybe not DNA *per se*, but at least genomic DNA can be seen as "the ultimate template of our heredity" (Jenuwein and Allis 2001, 1074). Also, the knowledge about different profiles of epigenetic gene expression does not automatically lead to abandon the dream of controlling life. For example, Pennisi (2001, 1067) talks about *future epigenomics* in which the epigenetic patterns regulating DNA will be easily determined, aberrant patterns will be linked to diseases, while their manipulation will lead to controlled changes of gene expression, thereby realizing the genomic dream at the epigenetic level. The question rises if we are merely witnessing an expansion of *gene-centrism* into *epi-gene-centrism*. Is epigenetics about turning a DNA-world into an RNA-world or an enzyme-world, in terms of looking for a new *unmoved mover* in the causal chain of life? Or why not hold on to the gene-centric view and argue that epigenetic phenomena are ultimately encapsulated in a genetic landscape as many of the enzymes involved in epigenetic processes are encoded anyway in the genes (cf. Morange, 2002). Likewise, one could argue that the qualification of epigenetics as a system of inheritance stands in virtue of the genetic inheritance system in which genes remain the ultimate controlling factor. Or argue that epigenetics merely sees development as a function of the cellular heredity of epigenetic control, i.e. it sees development *as* heredity. According to Griesemer (2002), this would make epigenetics into nothing but another remnant of the Weismannist perspective, absorbing epigenetics into the tradition of classical genetics. However, does this argumentation hold? And if not, how and where does the alternative side of epigenetics come through?

In order to answer this question – while at the same time illustrating the consensus definition – a closer look at the epigenetic practice is needed.

Epigenetic research indeed plays with molecular genes, but foremost, it plots them out in a dynamic four-dimensional time-space. As such, DNA – in form and function – is accounted for in its *in vivo* appearance, i.e. as differentially folded chromatin. Where before chromatin was considered as a passive result of automatic folding of DNA sequences, epigenetics acknowledges that the manner of genome packaging both influences gene expression (Kass and Wolffe, 1996), as well as it is influenced by non-DNA factors. For example, genomic imprinting by chromatin marks such as methylation[9] shows that the *same* DNA sequence can have several different epigenetic patterns or *epialleles*, "each pattern being

[9] DNA methyltransferases put methyl groups on the cytosine bases of CpG dinucleotides on both complementary DNA strands, thereby forming methylation patterns. These do not change the coding properties of the codons in which these bases participate, but induce the formation of heterochromatin leading to the inactivation of gene expression (Jones and Takai, 2001).

related to a different functional state" (Jablonka, 1994, 304). This leads to conclude, "whereas a highly complex assortment of genotypes may lead to a spectrum of phenotypes, the same spectrum might result if a single genotype generates a highly complex assortment of epigenotypes" (Whitelaw and Martin, 2001, 364). As these patterns can be induced by environmental or developmental stimuli, epigenetics extends the notion of *variation* and *mutation* beyond the classical view. Moreover, as alluded to in the consensus definition, the epigenetic status of a gene also extends the notion of *inheritance* as methylation patterns and their variations have been detected[10] to be heritable during mitotic and meiotic cell division in both vegetative and sexually reproducing organisms.[11] This type of inheritance differs from *genetic* inheritance by being less stable (epigenetic mutations are more prone to be reversible) and more sensitive to variations. Nonetheless, it allows "the inheritance not only of instructions, but also of the products of instructions, or a particular 'interpretation' of instructions" (Jablonka, 1994, 305). As such, chromatin marking systems not only enable cells or organisms to respond much more flexibly to environmental signals conveyed by hormones, growth factors, and other regulatory molecules *without having to alter the DNA itself* (Pennisi, 2001). They also allow to inherit the *functional state* of a DNA sequence.[12]

It has been discussed elsewhere (Van Speybroeck, 2000) how epigenetic research slowly uncovers specific factors and processes impinging on the genome and influencing how a specific DNA-sequence is used (within or across generations). Hereby, epigenetics implicitly introduces a *layered view on genomic contexts*, in which these contexts range from intracellular, cell-cellular, intercellular, to organismal and environmental influences. Supported by an impressive technological evolution (Petronis, 2001), the idea of layers of genomic contexts further explores Waddington's general notion of *context-dependency* (Van Speybroeck, 2002). In essence, it helps to reveal how DNA is not just embedded in 'a' complex environment that somehow is 'out there', but how this environment is intimately and specifically interwoven with gene action and function.

Regarding the status of this layered or stratified organization, it can be said that it minimally stands for a *heuristic* tool, a practical research strategy, in which reductionist means are used in a more encompassing, 'complex' setting. A stratificational viewpoint as understood here allows to incorporate reductionist strategies

---

[10] Based on the DNA replication mechanism and the existence of molecular machinery that recognizes hemimethylated sequences and converts them into a fully methylated state (Lewin, 1998).

[11] Heritable epimutations are often observed in case of transgene silencing in plants, but also in laboratory strains of *A. thaliana* and maize (Martienssen and Colot, 2001). For a study of the naturally occurring heritable epimutant of *Linaria vulgaris*, described more than 250 years ago by Linnaeus, see Cubas et al., 1999. By now epigenetic inheritance has been shown to exist in many species, including mammals (Jablonka and Lamb, 1995, 134–137, see also Roemer et al., 1997; Morgan et al., 1999).

[12] Jablonka (1994) argues that in fluctuating environments lasting longer than the lifespan of the organism, but shorter than the time usually required for the fixation of an advantageous genetic mutation, an epigenetic inheritance system may provide an adaptive potential by preventing physiological stasis and allowing fast formation of new variations.

in that it does individuate levels and subscribes to the existence of system boundaries. But it involves an ongoing conditional interpretation of these levels and boundaries in the sense that (i) what counts as levels and boundaries is defined from within specific conditions, and (ii) these conditions are subject to change and revision, not only through new experimental outcomes, but also through continuously evolving research interests and purposes. In delineating the specific material conditions at various layers, epigenetics therefore contributes to a more detailed articulation of what is often captured under the general heading of self-organization and complexity. As *in itself* a stratification of genomic contexts does not allow to favor any of these contexts, and as it is, amongst other things, the epigenetic *practice* that sets apart specific contexts, the scientist (as well as the philosopher) is here confronted with the task to make these choices explicit by clarifying the interests and purposes at play in each context, in its interrelatedness with other contexts. This not only makes epigenetics a rich environment for the generation of new hypotheses and research areas, it also enables to ask how far the assumed vagueness or fixity of boundaries in a biological system problematizes the nature of scientific explanation. From a philosophical point of view, indeed it becomes interesting not only to investigate how under experimental practice biological (sub)systems come into being or, *vice versa*, lose their integrity, but also when an explanation is considered adequate.

The epigenetic perspective no longer rests on a dogmatic framework, pinning down its results and views to one ultimate level of biological organization. This defies a relapse into the extremes of reductionism, but also into the extremes of holism, which *in extremis* not only makes any boundary between a system and its environment disappear (cf. Oyama, 1985), but also runs the risk of making any form of analysis *a priori* impossible as traditional biological boundary notions such as *nucleus*, *cell*, *tissue*, *organ*, *organism* are deprived of meaning. In comparison, epigenetics recognizes that these notions at any time remain relative to the research context, but acknowledges the usefulness of constructing provisional 'insides' and 'outsides' in an 'architecture of systems' (cf. Shapiro, 2002). As such, the organismal level need not be the most dominant source of influence on the genome at any time in any case, nor does one always need to take into account the entire whole. Likewise, one need not throw away some of the more valuable *genetic* insights, such as the idea that also the organismal level can be constrained by genetic characteristics. Epigenetics instead beholds the potential to investigate both the nature of these constraints and the conditions under which an organism's history and actuality influence its (heritable) genetic constitution. That these constraints depend on the specific organism studied is a further elaboration hereof.

Via the notion of *context*, a means is found to transcend a reductionist view on genes as sole organizers of both biological organisms *and* biological knowledge. Within an epigenetic framework, genes no longer stand for inviolable molecular atoms 'causing' the organism, but rather for temporarily relatively stable units which take form within a biological system, i.e. a dynamic self-organising system in which the partaking factors interpret one another, and through this interpretation construct each others functional meaning. As such, genes only come into being through a larger context. The idea of context here involves a nuancing of the classical distinction between ontological and epistemological claims, as system boundaries will only *be* in as far as they are revealed in and through a certain context. Hence, it comes as no surprise that the concept of what a gene *is*, has changed and continues to vary. As a *unit of heredity*, it cannot unequivocally be defined as a DNA-sequence because also non-DNA-variations (or epimutations) can achieve a heritable status. And as a *unit of development*, i.e. a functional unit of the regulation and expression of DNA, it cannot be singled out *a priori*. As such, if granted a physical identity in terms of DNA, genes simply cannot be without an immediate specification of their material and functional embedding. This promotes the idea that the term *gene* necessarily harbors a huge diversity in interpretations, demanding a flexibility of its users.

4. IMPLICATIONS FOR PHILOSOPHY OF BIOLOGY

In describing gene-centrism or even molecular biology as being "an ultimately reductionistic conquest" (Rheinberger, 2000, 231), philosophy of biology is challenged by the 'epigeneticization' of biological thinking to take a closer look at 'gene-talk' and to investigate *why* gene-centric discourses seem to keep their place within biology. Portin (1993, 208) argued that "our knowledge of the structure and function of the genetic material has outgrown the terminology traditionally used to describe it". But if gene-talk no longer serves to promote the gene-centric paradigm, and is used as a shortcut language to express complex knowledge, questions can be raised about both the terminological coherence of the alternative view and the need to (and possibility of) develop(ing) an appropriate vocabulary.

While residing under the term 'complexity', this alternative view makes ample use of terms like *emergence*, *autonomy*, and *(self-)organization*. It is interesting to note that already in the 1940s the developmental biologist Conrad H. Waddington remarked that "there has been a tendency either to regard *organisation* as one of the irreducible fundamental bases of all biology, or to invoke it, as though it were a well-defined concept, to fill up any awkward gaps in a theoretical structure" (Waddington, 1947, 143, italics added). But as Waddington argued, "the scope of the notion, as an *explanatory* principle, is not so great as has sometimes been suggested. On the other hand, it does provide a valuable *method of thought*" (ibid., 143, italics added). Does this mean that the abstract concept of 'organization' only serves the purpose of *indicating* or *presupposing*, perhaps of *generally describing*, the idea that parts behave differently within a totality? does it indicate certain usefulness at the general *exploratory* level, without necessarily offering *explanatory* power? And is it in the context of living systems indeed correct, and/or relevant, to subscribe to a distinction between exploring (interpreting, describing) and explaining?

As we have seen, epigenetic research has the potential to focus our thinking about contextual determination, by articulating the role of biological organization and organizational levels in the (relative) individuation and instantiation of certain entities. As such, it invites philosophers and scientists to look beyond the ideals set out by the classical scientific viewpoint, and to overcome the disappointments of not finding, as in physics, the general, abstract (categorical) terms to adequately describe living systems, or of not being able to formulate and predict the behaviour of living systems on the basis of general laws. Perhaps epigenetics can teach us that there is no intrinsic need to strive prominently for general theories and insights about the living, and that it is relevant to take up the genuine challenge of adequately interpreting and understanding the innovating experimental research results and methodologies available in the biological sciences. This is valuable, in as far as it can help scientists as well as philosophers to trace, to articulate and to make explicit the complex struggle of objectification, which, each time again, is a struggle to find those questions that capture most adequately the phenomena that are brought to the foreground within a specific context.

In this regard, the philosophical debate on the gene concept (cf. Falk, 2000; Morange, 2000, 2001) forms an interesting case study. Whereas this debate long focused on the evolution of the Mendelian gene concept into the classical molecular gene concept,[13] today attention is given to the diversity of possible interpretations when talking about genes[14]. A discussion has been launched on what the status is of these repeated genes, split genes, overlapping genes, assembled genes, jumping genes, nested genes, master genes, switch genes, selector genes, master control genes, key regulatory genes, developmental genes, developmental control genes, and so on.[15] Apparently, this diversity not only contrasts with the idea of a gene corresponding to a fixed point on the chromosomes or to a unique type of structure at the molecular level coinciding with a unique type of functionality. Within philosophy

---

[13] Portin (1993) analyses the *Mendelian gene* as being at the same time (i) a unit of genetic transmission from one generation to the next, (ii) a unit of recombination, (iii) a unit of mutation, (iv) a unit of development. The *molecular gene* is seen as (i) a unit of genetic function in terms of cistrons coding for a single polypeptide, (ii) a unit of mutation in that a change in one of a few nucleotides is enough to lead to an altered phenotype, (iii) a unit of recombination in terms of the smallest unit of genetic material that can be separated from other such units by genetic recombination, and (iv) a unit of transcription as depicted in the Central Dogma.

[14] In molecular biology, *a gene* can stand for "an undefined unit, a unit-character, a unit factor, a factor, an abstract point on a recombination map, a three-dimensional segment of an anaphase chromosome, a linear segment of an interphase chromosome, a sac of genomeres, a series of linear subgenes, a spherical unit defined by a target theory, a dynamic functional quantity of one specific unit, a pseudoallele, a specific chromosome segment subject to position effect, a rearrangement within a continuous chromosome molecule, a cistron within which fine structure can be demonstrated, and a linear segment of nucleic acid specifying a structural or regulatory product" (C.E. Axel in Nelkin and Lindee, 1995, 202).

[15] *Repeated genes* exist out of nucleotide sequence repetitions of DNA and are no unit of genetic transmission or of transcription. From *split genes* one single transcript (heterogeneous nuclear RNA) is produced from which the introns are removed during the processing of messenger RNA (mRNA). *Overlapping genes* share parts of the same DNA-sequence, but express different products. *Jumping genes* are bits of DNA that move throughout the genome by excision and insertion, often leading to mutations or chromosomal rearrangements and affecting gene expression. *Nested genes* reside within an intron of another gene (Portin, 1993). For the other terms, see Morange, 2000, 193.

of biology it also unleashed the ambition to 'disintegrate' (Beurton, 2000, 281) or 'deconstruct' the gene. An ambition that unfolds itself as (i) redefining the gene-concept,[16] (ii) abandoning the gene concept,[17] or (iii) devising a new terminology[18] in order to bring a halt to the so-called *confusion* in the biological landscape.

This confusion is mainly attributed to biologists. For example, Beurton (2000, 296, italics added) considers it important "that *geneticists* recognize the many levels at which genes can be perceived". However, as the growing consensus in this debate – labelled *the developmental gene concept*[19] (Falk, 2000, 331) – captures about the same concept already present in molecular biology[20] (i.e. genes do not pre-exist the processes by which a polypeptide comes into existence, but only become 'real' the moment of their contextualized functioning[21]), two issues become pertinent within philosophy of biology.

Firstly, it becomes important to explore and question the value of rigidly labelling concepts. Rheinberger (2000) here argues for the heuristic usefulness of the 'fuzziness' of the gene-concept and claims that it is "counterproductive, to try to sharpen the conceptual boundaries of vaguely bounded research objects while in operation" (Rheinberger, 2000, 221). In this respect, he even reminds that the rise of molecular biology has come about without a comprehensive, exact, and rigid definition of what a gene is. However, if the developmental gene concept has to

---

[16] Cf. the *molecular gene concept*, defined by Waters (1994, 178) as "a gene for a linear sequence in a product at some stage of genetic expression". Or *gene-P* and *gene-D*, respectively defined in Moss (2001, 88) as being an instrumental tool "in predicting a phenotypic outcome (…) most often based upon the absence of some normal sequence" and "a specific developmental resource, defined by its specific molecular sequence and thereby functional template capacity and yet it is indeterminate with respect to ultimate phenotypic outcomes". But also the *PMG* or *process molecular gene*, described in Neumann-Held and Rehmann-Sutter 1999 and in Neumann-Held, 2001 as "a process including any particular kind of (molecular) relation between DNA sequences, causal non-DNA inputs and mechanisms for the production of polypeptides".

[17] Philip Kitcher F. Ex and some DST-proponents (see Neumann-Held, 2001).

[18] Cf. the attempt of Brosius and Gould in 1992, introducing a *nuon* (any segment of DNA with recognizable structure and/or function), *potonuons* (duplication, amplification, recombination, retroposition of nuons), *naptonuons* (potonuons dissipating their nonadaptive former information without acquiring new ones) and *xaptonuons* (potonuons that exapted to a new function) (see Rheinberger, 2000).

[19] A developmental gene is interpreted in terms of "a dynamic function temporarily conferred to DNA stretches by the genome's reading apparatus for the purposes of producing a protein" (Beurton, 2000, 293). This concept however ignores the presence of genes coding for RNAs only. In this regard, Water's molecular gene concept (see note 15) is more encompassing.

[20] Morange (2000) illustrates how developmental and evolutionary biology from the 1980s onwards got familiar with this concept through the discovery of homeobox genes (which are strongly conserved throughout evolution, but control the development of very different organisms built along different pathways and bodyplans). Cf. also Fogle (in Falk, 2000, 331) who claims that "no self-respecting geneticist claims that DNA is solely required for a causal explanation of cellular processes".

[21] Although one could argue that genes pre-exist developmental processes via their evolutionary origin. However, here genes also are *products* (and not starting points) of evolution at the population level in that only "an array of nonlocalized DNA variations, whose differential reproduction comes to be controlled by some such adaptive difference large enough for selection to detect, begins to qualify as a gene" (Beurton in Neumann-Held and Rehmann-Sutter 1999, 90).

point out an essence or a universal adequacy, does one not also take the risk of ending up with a very general label that might well become trivial? Still, it can be of assistance in giving word to the 'epigeneticization' of biological thinking, especially in those areas, which have been isolated from the conceptual changes taking place in molecular biology. Of course, one could leave the communicative aspect aside and, instead of abstracting away the typical pluralism of how a gene can be interpreted biochemically, concentrate on asking "better questions about the relations between genes, environment and phenotype" (Gifford, 2000, 49). This is an extremely challenging route, especially because acknowledging this pluralism demands an interest for how 'genic contexts' come into being and can be studied practically. The grafting of the term 'gene' upon the experimental framework also is interesting because this allows to take notice of the communicative and pragmatic context in which biologists work.

Secondly, it is important to investigate exactly *for whom* the diversity of 'genic materialisations' is confusing. Morange points out that "scientists have always been much more comfortable with vague concepts than have philosophers. (...) a vague concept is often rich in explanatory potential" (Morange, 2001, 28). However, is the diversity regarding the gene concept really vague? And are *molecular biologists* working with vague descriptions when talking in terms of *genes*? The empirical study performed by the Stotz and Griffiths (2004) managed to give a refreshing input to these questions. Their survey aimed to bring clarity in how post doctoral biologists from diverse disciplines make use of gene concept(s) in specific communicative contexts.[22] Although the authors conclude somewhat quickly that the classical molecular gene concept (i.e. DNA coding for the primary structure of a protein) continues to function as a stereotype for biologists,[23] they also confirm that biologists manage to handle a huge amount of different specific gene concepts without apparent conflicting problems. In fact, biologists use a *methodological gene concept* in which a gene's identity is made relative to the context in which it is held to be active.[24] Determining what a gene is then is "a methodological process" (Fogle, 2000, 4) based on a specific research programme. As such, a continued

[22] All biologists were attached to the University of Sydney (Australia) and came from disciplines ranging from classical biology to biochemistry, agriculture, veterinary science, medicine and pharmacology). From the 250 scientists receiving the questionnaire, only 80 responded.

[23] It remains to be seen in how far this is a by-product of the survey or not.

[24] Cf. Rheinberger (2000): *biophysicists* working with a crystalline DNA fibre and an X-ray apparatus see genes as particular conformations of a double helix in which the atomic co-ordinates of the nucleic acid bases are important, *biochemists* working with isolated DNA fragments in test tube see them as nucleotide polymers exhibiting certain stereochemical features and sequence patterns, *molecular geneticist* see them as informational elements of chromosomes giving rise to specific functional or structural products, *evolutionary geneticists* see them as products of mutated, reshuffled, duplicated, transposed bits of DNA within a complex chromosomal environment evolved through differential production, selection, ..., *developmental geneticists* see genes as hierarchically ordered switches inducing differentiation when switched on/off or as patches of instructions that are realized in expression (here also the regulatory aspects of the genetic circuitry are important).

use of the classical molecular gene concept not necessarily denotes a stereotypic reference to gene-centrism, but can be demonstrative of a discourse in which biologists handle the necessary background knowledge that adds subtle differences in meaning to the words used.[25] In this regard, biologists not only jump flexibly between a classical-Mendelian and a molecular interpretation. Also differences in molecular interpretation[26] are grasped without much ado, while the complexity of all these possibilities is much harder to capture for any non-biologist. In other words, a biologist is familiar with what others might consider as 'details'.

In sum, while Rheinberger argues that biologists have understood how to benefit from 'fuzzy' concepts, here it is recognized that biologists use the gene concept methodologically and manage to handle a multitude of *precise* manners to fill in that concept. As a consequence, instead of trading "trying to codify precision of meaning" for "an epistemology of the vague and the exuberant" (Rheinberger, 2000, 222), it might be more interesting for philosophers to reconsider this 'codification of precision' and to explore in how far *precision* leaves room for *flexibility*.

5. CONCLUSION

Although the current discourse in molecular biology continues to use the gene-centric shortcut language of 'a gene for x', this use immediately is shaded because a DNA sequence – the (epi)genetic 'working unit' par excellence – may be *terminologically* or *materially* isolated for convenience, as soon as it is *experimentally* inserted or localized in a living cell or organism, its regulation and function only exist in virtue of a larger context. *Epistemologically*, this implies that the organization of biological knowledge no longer circles around the gene concept itself, but rather around the notion of 'larger contexts' in which a gene is characterized or even comes into being. The *ontological* consequence of this is that a gene – as a functional unit – ceases to stand on itself as a mere DNA-sequence, while *methodologically*, the crucial challenge of the epigenetic practice is to determine on the basis of detailed experimental research results, the precise meaning and scope of these genetic contexts. An important implication of these shifts is that a gene (what it is and how it functions) has changed from *explanans* to *explanandum*, i.e. from that which explains to that which is in need of an explanation.

This perspective, taken to its full range, unfolds as a dynamic systems approach (cf. Van de Vijver et al., 2003). It not only recognizes that in order to give meaning to 'a gene', the choice of any genic context is relative to the phenomenon

[25] Although Waters argues that there is "disagreement among biologists concerning the question of what is a gene" (1994, 164), he also states "molecular biologists understand what the term "gene" refers to in concrete situations because the context of discussion implicitly indicates the relevant stage and product of genetic expression" (1994, 179).

[26] E.g. does the gene in question contain merely the coding sequence, or also the promoter? And what about enhancers, silencers, methylation patterns, etc? Is it one or two genes we are talking about?

or system under investigation, but also that it is relative to the perspective the researcher adopts. Therefore, epigenetic research has the potential to lead to a more flexible view on the identification and description of biological entities, and is as such important because it can make philosophers and scientist aware of the intimate relationship between ontological and epistemological issues, e.g. between the eventual identification and individuation of certain entities, and the questions on the basis of which these entities acquired their status as meaningful entities. This flexibility already fuels the debate in philosophy of biology on the gene concept, by ... unifying definition is often much of a "rhetorical exercise" (Rheinberger, 2000, 255) risking to evade the genuine challenge related to contextuality.

ACKNOWLEDGEMENTS

The authors are indebted to the editors of the volume and to two anonymous referees for providing the opportunity to further shape this paper. L. Van Speybroeck especially thanks the Bijzonder Onderzoeksfonds (Ghent University) and the Fund for Scientific Research, FWO-Flanders, for financial support.

REFERENCES

Beurton PJ (2000) A unified view of the gene, or how to overcome reductionism. In: Beurton PJ, Falk R, Rheinberger H-J (eds) The concept of the gene in development and evolution. Historical and epistemological perspectives. Cambridge University Press, pp 286–316.

Bolker JA (1995) Model systems in developmental biology. BioEssays 17 (5):451–455.

Crick FHC (1958) On protein synthesis. Symp Soc Exp Biol XII:138–163.

Cubas P, Coral V, Coen E (1999) An epigenetic mutation responsible for natural variation in floral symmetry. Nature 401:157–161.

Dawkins R (1989) (1976) The selfish gene. New edition. Oxford University Press, Oxford.

Falk R (2000) The Gene – A concept in tension. In: Beurton PJ, Falk R, Rheinberger H-J (eds) The concept of the gene in development and evolution. Historical and epistemological perspectives. Cambridge University Press, pp 317–348.

Fogle T (2000) The dissolution of protein coding genes in molecular biology. In: Beurton PJ, Falk R, Rheinberger H-J (eds) The concept of the gene in development and evolution. Historical and epistemological perspectives. Cambridge University Press, pp 3–25.

Gifford F (2000) Gene concept and genetic concepts. In: Beurton PJ, Falk R, Reinberger H-J (eds) The concept of the gene in development and evolution. Historical and epistemological perspectives. Cambridge University Press, pp 40–68.

Gilbert SF (1996) Enzymatic adaptation and the entrance of molecular biology into embryology. In: Sarkar S (ed) The philosophy and history of molecular biology: new perspectives. Kluwer Academic Press, Dordrecht, pp 101–123.

Griesemer JR (2002) What is 'epi' about epigenetics? In: Van SL, Van de VG, De Waele D (eds) From epigenesis to epigenetics: the genome in context, vol 981, Annals of the New York Academy of Sciences, pp 97–110.

Henikoff S, Matzke MA (1997) Exploring and explaining epigenetic effects. Trends Genet 13(8):293–295.

Holliday R (1987) The inheritance of epigenetic defects. Science 238:163–170.

Holliday R (1994) Epigenetics: an overview. Dev Gen 15:453–457.

Holliday R (1996) DNA methylation in eukaryotes: 20 years on. In: Russo VEA, Martienssen RA, Riggs AD (eds). Epigenetic mechanisms of gene regulation. Cold Spring Harbor Laboratory Press, New York, pp 5–27.

Jablonka E (1994) Inheritance systems and the evolution of new levels of individuality. J Theor Biol 170:301–309.

Jablonka E, Lamb MJ (1995) Epigenetic inheritance and evolution: the lamarckian dimension. Oxford University Press, Oxford.

Jablonka E, Lamb MJ (2002) The changing concept of epigenetics. In: Van SL, Van de Vijver G, De Waele D (eds) From epigenesis to epigenetics: the genome in context, vol 981, Annals of the New York Academy of Sciences, pp 82–96.

Jenuwein T, Allis CD (2001) Translating the histone code. Science 293:1074–1080.

Jones PA, Takai D (2001) The role of DNA methylation in mammalian epigenetics. Science 293:1068–1070.

Kass SU, Wolffe AP (1996) Histones, Histone modification, and the inheritance of chromatin structure. In: Russo VEA, Martienssen RA, Riggs AD (eds) Epigenetic mechanisms of gene regulation. Cold Spring Harbor Laboratory Press, New York, pp 529–545.

Keller EF (1995) Refiguring life: metaphors of twentieth-century biology. Columbia University Press, New York.

Lewin B (1998) The mystique of epigenetics. Cell 93:301–303.

Lewontin RC (2001) Gene, organism and environment: a new introduction. In: Oyama S, Griffiths PE, Gray RD (eds) Cycles of contingency. developmental systems and evolution. The MIT Press, Cambridge, pp 55–58.

Martienssen RA, Colot V (2001) DNA methylation and epigenetic inheritance in plants and filamentous fungi. Science 293:1070–1074.

Morange M (1998) (1994) A history of molecular biology. Original title: Histoire de la biologie moléculaire. Trans by Cobb M. Harvard University Press, Cambridge.

Morange M (2000) The developmental gene concept: history and limits. In: Beurton PJ, Falk R, Rheinberger H-J (eds) The concept of the gene in development and evolution. Historical and epistemological perspectives. Cambridge University Press, pp 193–218.

Morange M (2001) The misunderstood gene. Transl by Cobb M from La Part des Gènes, 1998, editions Odile Jacob. Harvard University Press, Cambridge.

Morange M (2002) The relations between genetics and epigenetics: a historical point of view. In: Van SL, Van de Vijver G, De Waele D (eds) From epigenesis to epigenetics: The genome in context, vol 981, Annals of the New York Academy of Sciences, pp 50–60.

Morgan HD, Sutherland HG, Martin DI, Whitelaw E (1999) Epigenetic inheritance at the agouti locus in mouse. Nat Genet 23:314–318.

Moss L (2001) Deconstructing the gene and reconstructing molecular developmental systems. In: Oyama S, Griffiths PE, Gray RD (eds). Cycles of contingency. Developmental systems and evolution. The MIT Press, Cambridge pp 85–98.

Moss L (2003) What genes can't do. MIT Press, Cambridge, MA.

Nelkin D, Lindee MS (1995) The DNA mystique. The gene as a cultural icon. W H Freeman and Company, New York.

Neumann-Held EM (2001) Let's talk about genes: the process molecular gene concept and its context. In: Oyama S, Griffiths PE, Gray RD (eds) Cycles of contingency. Developmental systems and evolution. The MIT Press, Cambridge, pp 69–84.

Neumann-Held EM, Rehmann-SC (1999). Individuation and reality of genes. A comment to Peter J. Beurton's article: 'Was sind Gene heute?' Theory Biosci 118:85–95.

Oyama S (1985) The ontogeny of information. Cambridge University Press, Cambridge, London.

Oyama S, Griffiths PE, Gray RD (eds) (2001) Cycles of contingency. Developmental systems and evolution. The MIT Press, Cambridge.

Pennsi E (2001) Behind the scenes of gene expression. Science 293 (Special issue on epigenetics):1064–1067.

Petronis A (2001) Human morbid genetics revisited: relevance of epigenetics. Trends Genet 17(3):142–146.

Portin P (1993) The concept of the gene: short history and present status. The Q Rev Biol 68(2):173–223.

Rheinberger H-J (2000) Gene Concepts: fragments from the perspective of molecular biology. In: Beurton PJ, Falk R, Rheinberger H-J (eds) The concept of the gene in development and evolution. Historical and epistemological perspectives. Cambridge University Press, pp 219–239.

Riggs AD, Martienssen RA, Russo VEA (eds) (1996) Epigenetic mechanisms of gene regulation. Cold Spring Harbor Laboratory Press, New York.

Robinson JD (1992) Aims and achievements of the reductionist approach in biochemistry/molecular biology/cell biology: a response to Kincaid. Philos Sci 59:465–470.

Roemer I, Reif W, Dean W, Klose J (1997) Epigenetic inheritance in the mouse. Curr Biol 7:277–280.

Ruiz-Mirazo K, Etxeberria A, Moreno A, Ibáñez J (2000) Organisms and their place in biology. Theory Biosci 119:209–233.

Sarkar S (1996) Biological information: a skeptical look at some central dogmas of molecular biology. In: Sarkar S (ed) The philosophy and history of molecular biology: New perspectives. Kluwer Academic Press, Dordrecht, pp 187–231.

Sarkar S (1998) Genetics and reductionism. Cambridge University Press, Cambridge.

Shapiro JA (2002) Genome organization and reorganization in evolution: formatting for computation and function. In: Van SL, Van de Vijver G, De Waele D (eds) From epigenesis to epigenetics: the genome in context, vol 981, Annals of the New York Academy of Sciences, pp 111–134.

Stotz K, Griffiths PE, Knight R (2004) How biologists conceptualize genes: an empirical study. Studies in History and Philosophy of Biological and Biomedical Sciences 35(4):647–673.

Thieffry D (1998) Forty years under the central dogma. Trends Biochem Sci 23:312–316.

Torres JM (1999) On the falsification of the *Central Dogma* and the *De novo* synthesis of molecular species. Philos Nat 36(1):1–18.

Van de Vijver G, Van Speybroeck L, Vandevyvere W (2003). Reflections on complexity in biological systems: Kant and beyond. Acta Biotheor 51(2):101–140.

Van Speybroeck L (2000) The organism: a crucial genomic context in molecular epigenetics? Theory in Biosci 119:187–208.

Van Speybroeck L (2002) From Epigenesis to Epigenetics: the case of CH Waddington. In: Van SL, Van de Vijver G, De Waele D (eds) From epigenesis to epigenetics: the genome in context, vol 981, Annals of the New York Academy of Sciences, pp 61–82.

Van Speybroeck L, Van de Vijver G, De Waele D (2002) Theories in early embryology. Close connections between epigenesis, preformationis, and self-organization. In: Van SL, Van de Vijver G, De Waele D (eds) From Epigenesis to Epigenetics: the genome in context, vol. 981, Annals of the New York Academy of Sciences, pp 7–50.

Waddington CH (1947) (1940) Organisers and genes. Cambridge University Press, Cambridge.

Waddington CH (1975) The evolution of an evolutionist. Cornell University Press, Ithaca, New York.

Waters KC (1994) Genes made molecular. Philos Sci 61:163–185.

Waters KC (2004) What was classical genetics? Studies in History and Philos Sci 35:783–808.

Whitelaw E, Martin DIK (2001) Retrotransposons as epigenetic mediators of phenotypic variation in mammals. Nat Genet 27:361–365.

Wu CT, Morris JR (2001) Genes, genetics, and epigenetics: a correspondence. Science 293:1103–1105; Full text at www.sciencemag.org/cgi/content/full/293/5532/1103/DC1.

PART III

GENETICS AND THE ETHICAL, LEGAL AND SOCIOLOGICAL DEBATE

CHAPTER 9

IS DNA REVOLUTIONIZING MEDICINE?

ANNE FAGOT-LARGEAULT
Chair of Philosophy of Biological and Medical Sciences
Collège de France, Paris, France
anne.fagot-largeault@college-de-france.fr

1. INTRODUCTION

The course of the discovery of the DNA molecule's double-helix structure (*Nature*, 1953, 171: 737–738) has recently been reassessed by historians (Abir-am, 2003), as the fiftieth anniversary of the discovery was celebrated in 2003 with great acclaim (*Nature*, 2003, 421, 395–453); but did the discovery really revolutionize medicine? As far as clinicians can see, up to now molecular genetics has had little impact on medical practice, and the question is still asked using the future tense: "Will genetics revolutionise medicine?" (Holtzman and Marteau, 2000). However, some sociologists and philosophers of medicine have made it clear that they are worried about the devastating consequences that genetic knowledge, genetic tests, and biotechnologies, could have on what is usually known as medical humanism. This is not the first time that technoscientific progress has been perceived as threatening an "anthropological collapse". Writing about the introduction of the practice of autopsy into medical schools at the end of the Middle Ages, Le Breton wrote: "Modern medicine is indeed an effect of this anthropological collapse. It is no longer Man that is its concern in his history and his person, but disease and the body viewed as a machine" (1993, 127). It treats people as things: something of which the whole of western scientific culture has been accused. Edmund Husserl deplored that philosophy has not been able to prevent science from sliding down this slippery slope: "The ultimate meaning of the accusation that should be addressed to

[1] Translated into English by Monika Ghosh, revised by the author.

A. Fagot-Largeault et al. (eds.), The Influence of Genetics on Contemporary Thinking, 137–150.

philosophy in general – with the exception of idealism – is that it has not been able to overcome naturalistic objectivism..." (1954, §73). Curiously the introduction of digital imaging into the field of medicine, in the 1970s, followed by that of robotics, has triggered another accusation (diametrically opposed to the previous one): whereas concrete knowledge of the body is essentially the sphere of the physician who palpates, taps, listens, and even more so that of the surgeon, who opens up the body, perceives the movements of the viscera, and has the life of this body beneath his or her fingertips, it is obvious that medical students now learn about the body from computer images; and that in ophthalmology, neurosurgery, urology, the practitioner often operates on an image; or even allows a robot to operate on the basis of digital data. As a result, actual contact with the physical body is lost; the medical practitioner operates on a diagram of the body based on abstract data. This has led to the hypothesis that the real danger of technoscience is less in the *reification* of the body than in its *intellectualization*... In any case, regardless of whether genetic information involves a *molecule* or a *concept*, the question raised here is that of the impact of genetic knowledge on medicine.

2. THE CLINICIAN'S VIEW: A MORE RESERVED ASSESSMENT

The publishers of *Nature* asked the clinician John Bell to assess this impact for the fiftieth anniversary issue of the journal. If asked what has really *revolutionized* medical practice during the second half of the 20th century, Bell says, we physicians think of the discovery of penicillin and its development by Chain and Florey (*Lancet*, 1941), or of the demonstration of Doll and Hill (*British Medical Journal*, 1950) of the causal link between smoking and lung cancer. These discoveries have saved hundreds of thousands of lives by making it possible to treat or prevent serious illnesses. The theoretical interest for medicine of the discovery of the molecular structure of the gene was all the greater because it had long been known that many human diseases have a hereditary component. However, from a distance of 50 years, apart from monogenic diseases that display Mendelian transmission and affect only a tiny fraction of the human population, doctors are still rather sceptical about the value of trying to identify genetic factors at all cost, since they realize that in the etiology of most diseases, the supposed genetic predispositions are entangled with epigenetic and environmental factors, and we are far from having unravelled them: "The reductionism that accompanies molecular genetics will identify the pieces in the jigsaw, but assembling these to understand how complex systems malfunction will require a substantially more integrated approach than is available at present" (Bell, 2003, 416).

Pierre Corvol (Georges Pompidou European Hospital and Collège de France), who specializes in cardiovascular diseases, is of the same opinion: molecular genetics has raised great hopes and enthusiasm, people thought that it was going to revolutionize medicine, and "we went from illusion to disillusion". Today, even though we can indeed see its importance for *theoretical knowledge*, we have to note that it has not changed the approach to dealing with hypertension, for instance.

A rapid survey of medical specialties confirms that for monogenic diseases, which are rare diseases, genetic knowledge really has changed things; however, it has only been a *diagnostic revolution*. Let us take the example of Huntington's chorea, a disease with dominant autosomic transmission, characterized by a rapidly fatal degeneration of the nervous system, which reveals itself in adults, at an age at which patients may already have transmitted it to their offspring (with a 1 in 2 probability for each birth). Till recently, families carrying this disease had no way of avoiding it other than not having children. Since the 1980s a genetic test has been available, with the possibility of a prenatal diagnosis that can be followed by a pregnancy termination if the test proves positive. Associations of affected families welcomed this test (see World Federation of Neurology, 1989), particularly as there is no validated treatment for the disease once it has revealed its presence. Cell therapy trials are in progress, using transplants of neurons harvested from aborted foetuses, but these attempts are still at the experimental stage.

Apart from genetic counselling, there is one field in which the technologies derived from molecular genetics have led to a minor, but real, revolution: this is that of infectious diseases. The possibility of carrying out a laboratory diagnosis of a parasite, a bacterium or a virus using RT-PCR[2] (without having to wait for a culture), has made it possible to save time, and this can be crucial for the patients.

For cancer, rheumatic diseases, and diseases of the immune system, genetic methods are beginning to contribute diagnostic refinements or predictive factors. For instance, in paediatric oncology, amplification of the *n-myc* gene makes it possible to determine the aggressiveness of a glioblastoma; in rheumatology, the presence of factor HLA B 27 is known to increase the relative risk of rheumatoid arthritis 50 fold. In these disciplines, genetic information yields its main contribution by increasing the precision of classifications. Patient management has not changed radically.

The author of this paper has had the opportunity to work with a group of psychiatrists specialising in genetic research in the field of psychiatric disorders (see Leboyer and Bellivier, 2003). In addition to animal research, this team has been monitoring two cohorts of patients with bipolar disease and schizophrenia (and their families) for some years now, the aim being to identify predisposing genetic factors in these diseases. The same team has joined with an international consortium carrying out research into the molecular genetics of autism: some of their findings in this field have been published. Mutations have been identified, that could have links with autistic disorders (see International Molecular genetic Study of Autism Consortium, 1998; Jamain Stéphane et al., 2003). What impact have these studies had on patient follow-up, and the advice given to families? A very meagre impact. The only molecular diagnosis that is sometimes sought in psychiatric practice, in the context of mental retardation or atypical autism, is that for fragile-X. Here genetic counselling can be contributive, and a prenatal diagnosis of a fragile X is available.

---

[2] PCR: polymerase chain reaction. RT-PCR uses a reverse transcriptase (RT) enzyme.

However, deciding whether to terminate a pregnancy on the basis of the test is not straightforward: mental retardation is expressed by only 80% of the boys and 30% of the girls carrying the mutation. We are there on the fringes of psychiatry: mental retardation is not a mental illness. The 2001 *World Health Report*, published by the World Health Organization (WHO, 2001), shows that mental disorders weigh very heavily on health systems. For most of the major psychiatric diseases (schizophrenia, mood disorders, obsessive-compulsive disorder, addictions), it is assumed that genetic factors for vulnerability must exist; but we know that they interact with epigenetic and environmental factors. Briefly, psychiatric diseases are multifactorial diseases, like diabetes, hypertension, obesity, senile deafness, etc. It is not therefore reasonable to imagine that they could be eradicated, because the gene pool consists of the entire general population (which puts paid to the ancient eugenic obsession that the problem could be dealt with by preventing patients from reproducing). What we can hope for from genetic research is the further subdivision of existing entities, for example, the ability to distinguish between several types of schizophrenia, making it possible to introduce appropriate adjustments of treatments (although this is still a *hope*, and not yet a reality). Apart from this hope, the only real benefit that has been obtained from progress in genetics is that it has contributed to the greater *social acceptance* of mental disorders. This is very obvious in the case of autism. Following the work of Bruno Bettelheim and the schools of psychoanalysis, many psychiatrists had taken on board the hypothesis that autistic disorders are of psychogenetic origin: the cause of the disorders manifested by the child were attributable to his or her parents' approach to child raising, and in particular to coldness on the part of the mother. This accounts for the treatment strategy that was proposed at the time – to separate the children from their parents, send the mothers for psychoanalysis, and help the children by placing them in a supportive and warm institutional context. Now that we know that a person is *born* autistic, and that the disorder is genetic or epigenetic in origin, the families have been freed from their burden of guilt, and involved in the educative approach. Autism has been tamed.

This is one result. However, one has to admit that the discovery of chlorpromazine (Laborit et al., 1952; Delay and Deniker, 1952), which was roughly contemporaneous with that of the double helix, albeit completely independently of it, has *revolutionized* the management of mental disorders to a considerably greater extent than any aspect of molecular genetics.

3. POLEMICS ABOUT THE "GENETICIZATION" OF MEDICINE

Nevertheless, the suspicion has been aroused that even if modern genetics has not had any *direct* impact on the treatment of diabetes, hypertension or schizophrenia, it has *indirectly distorted* the spirit of medicine by promoting a dehumanising reductionism, and finding molecular causes, at the expense of providing global care for sick people. Two camps are at loggerheads about this conjecture or thesis that there is a perverse "geneticization" of medicine (and of Western culture). Philosophers and sociologists with a "constructivist" tendency have argued in favor of this thesis. Some historians of science are adamantly opposed to it. Underlying the controversy, what is at stake (amongst other things) is the position of medicine, between the natural sciences (*Naturwissenschaften*) and the sciences of the mind or spirit (*Geisteswissenschaften*).

According to historians, medicine officially became molecular in 1949 with the publication in *Science* of the article of Linus Pauling et al. (1949), entitled 'Sickle cell anaemia, a molecular disease'. This event was soon being described as a *paradigm shift* (a *scientific revolution*), and the story has since become the stuff of legend. In 1945 William Castle (a clinician) and Linus Pauling (a chemist), were sharing a compartment in a train taking them back to Chicago from Denver. They were discussing their work, and Castle spoke of having seen cases of sickle-cell anaemia amongst the American Black population. One of the distinguishing characteristics of these anaemias is that when the red cells are examined under the microscope, they appear to be sickle-shaped in a de-oxygenated medium. The clinician reckoned that his fellow chemist might be interested in the detail of the birefringence of these sickle cells. According to legend, Pauling then had a brilliant flash of intuition: haemoglobin is a protein, and if the haemoglobin is abnormal, then this must be due to a mutated gene – the disease has a molecular cause, the explanation lies in the gene! Two historians of science, Feldman and Tauber (1997), have scrutinized this pretty story. They have shown that Pauling reworked his account of the event later, to make it fit in with the legend. In reality the discovery took him four years (1945–1949). When he got back to Chicago, he left it to a student to try to work out the link between the chemical structure and the sickle shape; the student worked on the subject throughout 1946 without coming up with anything. It was only in 1948, after Wells and Singer had joined the team, that they had the idea of using the method of electrophoresis, which led to the landmark publication of 1949.

While carefully reconstructing the history of research into this disease, Feldman and Tauber have shown that sickle-cell anaemia (later to be known as drepanocytosis), which had been identified in 1910 on the basis of clinical criteria (anaemia with leg ulcers) with the description of the first case, was already viewed as a hereditary disease in 1922 after only three cases had been diagnosed (it was thought to exhibit dominant Mendelian transmission; in fact it is recessive); and that in the 1920s it gave rise to constant exchanges between laboratory research and clinical research. They conclude that there had been no take over of clinical research by molecular (chemical) research, but rather complementarity and collaboration.

However sociologists and some philosophers do believe that in the invasion of clinical medicine by genetics they can see a revolution that is all the more dangerous for being stealthy: this is the thesis of (creeping) *geneticization.*

The word "geneticization" was launched in 1992 by the Canadian sociologist Abby Lippman (1991), to denounce the risk of discrimination and of increased inequalities implied by the use of genetic tests. If the police, insurance companies and employers, start keeping genetic files, and if as a result of prenatal diagnosis foetuses are eliminated because of a genetic defect, on the one hand, human diversity is reduced to molecular diversity (we are no longer anything more than packs of molecules!), and on the other, people are threatened with tests for genetic normality. In fact the revolt against being coerced by *genetically correct* standards has its source further back, in the libertarian movement of the 1970s, which in the voice of Ivan Illich (1977) condemned the medicalization of society as leading to an obligation to conform: it is your duty to be vaccinated, to have a chest X-ray in order to get a job, to get treatment for your high blood pressure, ... and you lose the right to make decisions about your own health. Illich, as we know, claimed the right for individuals to refuse medical treatment, to rebel against the obligation to look after their health (not to smoke, etc), and he went as far as to question whether young diabetics should have insulin treatment forced upon them (one does have to admit, he did have the courage of his convictions: he did not seek treatment for the cancer that killed him).

Following Lippman, other sociologists in North America (such as Nelkin and Lindee, 1995, or, to a lesser extent, Renée Fox) and philosophers, including some in Europe (such as Henk ten Have), rile against the human genome project, the molecular genetics community, physicians who provide genetic counselling, and journalists who trumpet every minor discovery. The alarm is being raised: *western culture is being geneticized.* Genetics is no longer just a science, it has become an ideology. It has put into people's minds fantasies of artificial life, and hopes of having perfect children to order. Medicine is the accomplice of a new form of eugenics (*the gene police*). The attempt to root out bad genes makes the life of handicapped people who have the very genetic anomalies that are being screened for, miserable and painful. As an example of this cultural distortion, ten Have refers to an article by Jordan, which appeared in 1997 in *JAMA*, the Journal of the American Medical Association. The article mentioned a genetic anomaly associated with an ex-boxers' disease (*dementia pugilistica*). As soon as this fact was realized, this led to the injunction: "we must test all would-be boxers and prevent those who have this genetic characteristic from taking it up".

At the end of the 1990s the American *Journal of Medicine and Philosophy* and its European counterpart *Medicine, Health Care and Philosophy* were the forum for a major debate for and against the thesis of geneticization.

A historian of science, Adam Hedgecoe (1998), on the basis of opinion surveys carried out by C. Condit and M. Williams (1997) – on the European side we could cite the *Eurobarometer on Biotechnologies*, published by the European Union–, claimed that the fears expressed had no basis in fact. The panic about genetic testing by employers, insurers or the police, were based on expectations that had proved false, since after several decades there had been virtually no example of such abuse. The journalists responsible for communicating scientific information apparently didn't do such a bad job, since citizens have taken on board the idea that the influence of genes lies in the realm of probabilities. Genetic counselling has been given prudently, respecting the rights of individuals to make their own decisions, and the families who have received it have not complained. The conviction that the availability of a prenatal diagnosis for certain disorders (trisomies, fragile-X) was going to lead to criticism of the people it affects, making their life impossible, and making parents who do not want to terminate a pregnancy feel guilty, seems to be contradicted by the facts: when fewer handicapped children are born, they are made more welcome. Finally it should be noted that the 'ethnic cleansing' episodes that the late 20th century regrettably experienced were not founded on any medical or scientific grounds.

Henk ten Have rejects the objection that his warnings are not based on objective facts. Some of his statements recall those of Hans Jonas, but he bases himself mainly on Michel Foucault. The concept of "geneticization", he says, is a heuristic concept, and as such, it has a function that is not descriptive, but prospective. The philosopher has duty to *imagine the worst*, to see what it is that we have to avoid. It is good and helpful to *frighten oneself* by imagining the worst, in order to make sure that one does not end up there. The identification of a human being with his or her genome is an insult to human freedom. The concept that you are what your genes are, abandons you to your fate, in the same way as does the idea of the predestination of the soul in Christian theology: genetic essentialism eliminates personal responsibility. The professionals involved in genetic counselling claim of course that they are simply giving you information, and that you are free to decide for yourself. But once pharmaceutical companies market kits for genetic tests, as they are beginning to do, this is an incitation to use them, and despite yourself you are subjected to social pressure: you are no longer free. What is unusual in ten Have's position, is that he claims that his ideas are totally immune to the facts, suggesting that unlike the natural sciences, in which the hypotheses advanced by scientists are subjected to validation (or refutation) by experiment, the social sciences and humanities allow their practitioners to produce irrefutable hypotheses: "The geneticization thesis is developed in the humanities, cultural sciences and philosophy, and it introduces the perspectives of these disciplines into the debate on genetics which is mainly in the area of the natural sciences [...] In philosophical discourse only a few examples will suffice to make a specific point plausible. Sometimes, even examples in the world are not relevant since philosophy may concern itself with hypothetical thought-experiments to understand phenomena. [...] It is a misunderstanding of a philosophical thesis when notions and explanations from philosophical discourse are tested with the instruments and methods from the empirical sciences" (ten Have, 2001, 298).

Hedgecoe retorts dryly that pretending to frighten oneself may attract media attention, but does not benefit thought; that there is indeed a need to think about the possible consequences of integrating the technologies of molecular genetics into medicine; but that constructive reflection must be based on an accurate knowledge of the facts, that is, on field studies that make it possible to pinpoint where the real problems lie. He totally rejects the supposed immunity of the human sciences from contradiction by the facts: "ten Have seems to assume that if you require empirical proof, then you must be operating from within the framework of the natural sciences. This is plainly mistaken, as any historian or archaeologist could tell you. If a claim is made in, for example, history, then the historian is expected to back it up with evidence... This does not mean that the historian has succumbed to the natural sciences. [...] Within the social sciences, there are huge resources available to philosophers interested in exploring the process of geneticization, resources that will allow them to test their concepts and predictions about how this process occurs, but also to revise their assumptions in the light of empirical evidence" (Hedgecoe, 2001, 306–308).

Empirical evidence? – but look at what happened in Cyprus, then in Great Britain and in Canada, ten Have answers, referring to the study that he carried out with Hoedemaekers (1998), into genetic screening for beta-thalassaemia. Beta-thalassaemia is a disease affecting haemoglobin (a protein in the blood), due to the synthesis of an incomplete beta chain. This is a disease that follows a recessive autosomal transmission pattern, and it is common in the Mediterranean basin. Heterozygotes are asymptomatic. Homozygotes suffer from anaemia which, unless it is treated, leads to early death during childhood; with treatment (blood transfusions, the chelation of iron) life expectancy can be prolonged (up to 15 or 20 years) but with a miserable quality of life. In Cyprus, before the events of 1980, the prevalence of the disease was 15%: and this was a very heavy burden on this island population. A screening programme, linked to a massive information campaign, mobilized health-care professionals, teachers, political leaders, the Orthodox Church. The first phase was that of screening for heterozygotes: it was explained to people that a marriage between two heterozygotes meant that each pregnancy carried a one in four chance that the baby would have the disease (homozygote); and a pre-marriage screening certificate became mandatory. In the second phase, married heterozygotic couples were offered the possibility of a prenatal diagnosis, and a termination of pregnancy if the unborn child was identified as being homozygotic.

From 1982, virtually no more homozygotic children were born in Cyprus. It is important to note that this strategy has not in any way eradicated the gene (which is dispersed throughout the heterozygotic population), and that it has allowed heterozygotic couples to have children. We should also point out that the success of this strategy is based on informing the citizens, and that this will have to be continued for future generations. Should we see in this experience a disastrous example of the geneticization of medicine and culture, or an example of the intelligent and responsible use of genetic knowledge and the resources of medicine to serve responsible procreation? Henk ten Have opts for the former interpretation, whereas I myself tend to favor the second. Great Britain and Canada, both of which have ethnic groups affected by thalassaemia, have followed the example of Cyprus.

The termination of a pregnancy following the detection in the foetus of a severe genetic deficiency is indeed a serious moral decision that has to be weighed carefully. Tinkering with procreation may cause still greater problems.

4. CONCERNS RELATED TO THE DEVELOPMENT OF "REPROGENETICS"

In his book on biotechnologies, Claude Debru (2003) provides an excellent historical account of genetic engineering: the beginnings (1972), the realization of the risks involved in research, the moratorium and the safety measures resulting from the

Asilomar process, the early successes – the synthesis of human insulin by a genetically-modified bacterium (and then a yeast) – (1978), the creation of transgenic plants (1980) and animals (1981), then human gene therapy trials (some inept or premature, indeed). According to an American report published in 2003 (see Parens and Knowles, 2003), what we should now be worrying about is the transfer of the technologies derived from genetic engineering to medically-assisted human procreation.

The idea that the human species will one day apply to itself the same technological tools as it has used to modify plants or animals is not new. Jean Rostand wrote in 1950 (92): "whether it is by the genes in the nucleus or by the genes in the cytoplasm, it looks as though Man may end up producing major structural improvements in the human body". Erik Parens and Lori Knowles, the authors of the American report, start from an observation. During the second half of the 20th century, medically-assisted procreation (MAP) and molecular genetic research developed separately. MAP, with the aim of treating sterile couples, invented its own methods (IVF, ICSI,[3] etc.). Molecular genetics used bacteria, before launching itself into the investigation of more complex organisms, such as *Arabidopsis thaliana*, the mouse-ear cress, a small cruciferaceous plant, in order to map their genes (21000 genes, for Arabidopsis), and to identify the function of these genes. During the 1990s, the techniques developed by genomics entered the world of MAP. The word "reprogenetics" made its appearance in 1999.

Parens and Lowes have identified in recent publications some examples of the application of genetic technology to human procreation that cause them concern. One method that can be used to sort sperm on the basis of the weight of their DNA[4] makes it possible to select the sex: 430 children whose sex has been selected by their parents have been born after their father's sperm had been sorted using this method (see Fugger et al., 1998). Little Molly Nash suffered from Fanconi anaemia, her parents decided that in order to help her they would have a 'baby-doctor'. A pre-implantation genetic diagnosis was carried out of their embryos obtained *in vitro*. The embryo for reimplantation was chosen so that it would provide Molly with a younger brother or sister who would be both histocompatible and free of the disease (see Verlinsky et al., 2001). Molly was to be given a blood transfusion from the umbilical cord (the baby was born in August 2000). A third example: about twenty women whose sterility was linked to an anomaly of their mitochondrial DNA[5] were able to conceive after their oocytes had undergone a transfer of cytoplasm from the oocytes of donor women who were not suffering from this disorder (see Cohen et al., 1998; Barritt et al., 2000, 2001; Templeton, 2002). The

[3] IVF: *in vitro* fertilization; ICSI: intracytoplasmic sperm injection.

[4] Deoxyribonucleic acid.

[5] Mitochondria are organelles present in the cytoplasm of the cell. They contain different genes from those in the cell nucleus. The transmission of the mitochondrial genome is solely via the mother (in animals); the flagellum of the spermatozoon (which contains the mitochondria of the father) does not get into the oocyte during fertilization.

children born as a result of this procedure were 'apparently' normal (although two of the foetuses obtained did suffer from Turner's syndrome;[6] it is not possible to say whether this was merely coincidence).

There are innumerable ways of tinkering with human reproduction in this way. Some authors do not hesitate to look forward to the genetic "enhancement" of our species (Gordon, 1999), by means of correcting reparable genetic defects at an early embryonic stage.[7] Others fear deterioration, knowing that, for instance, ICSI allows a sterile father to have a male child who will also be sterile. Many worries have been expressed about ICSI (see American Society for Reproductive Medicine, 2000; Hansen et al., 2002). The concerns expressed by Parens and Lowes can be accounted for to some extent by the North-American context. In the United States the public debate has long been centred on the question of the right to abortion, the battle lines drawn between the *pro-life* and the *pro-choice* camps remain intact, and it has therefore been decided that nothing involving research on the human embryo or the creation of embryos for research purposes may be funded by public money. Consequently both MAP and embryo research have slipped into the private sector, with no state funding or control, and full freedom. The scientists who wanted to treat defective oocytes by transfusing cytoplasm were under no obligation to submit their protocol or to seek regulatory approval before proceeding unless the journals where they wanted to publish had specific ethical requirements. Parens and Lowes condemn this situation, and the alternative model they suggest is the British system, which is centralized, regulated and supervised, but at the same time, "one of the most liberal in the world". The British (and in their wake, the Swedes and the Canadians) have been able to define a *policy* on reprogenetics, and this policy has been a subject of public debate. At present modifications of the germ cell line are banned, as are gender selection for non-medical reasons and the creation of human-animal chimeras (all things that can be done legally in the USA within the private sector, where nothing is banned). Apart from these restrictions, British scientists are free to propose any experimental procedure they wish, including the creation of human embryo cells by nuclear transfer (for "therapeutic cloning"), but they have to persuade the regulatory authority that what they propose doing is of real scientific value. This flexible control provides protection against possible abuses.

Parens and Lowes fear that abuses may increase in the United States now that scientists have access to human embryonic stem cell lines derived from "spare"

---

[6] Individuals suffering from Turner syndrome are phenotypically women, small in size, and have only one X chromosome (instead of two). The XO genotype is not always compatible with life, but it is a fairly frequent cause of miscarriage (5%).

[7] In this context the threat has been voiced of a 'loss of genetic diversity', which is potentially dangerous for our species. Michel Morange rejects this argument: "The idea that genetic diversity is always a good in itself and that any manipulation intended to reduce this diversity, even to a limited extent, is bad, is highly contestable. Let us add that it will doubtless take several centuries, even with all world's biologists, to curb this genetic diversity as much as colonization or wars sometimes did within a few months in past centuries" (1998, 204).

human embryos left over from assisted procreation: which is ideal material for trying out genetic recombinations, and transform MAP into something like using a Meccano set, so that children manufactured to order could become consumer goods like any other. However, the problem extends far beyond the American context. It is a fact that MAP has opened a huge experimental field to genetic engineering. The boundary between human experimentation and clinical innovation has been blurred, it is accepted that in this field procedures may be tried out without previous animal testing, and there often is no reliable monitoring of outcomes. Couples who resort to these procedures tend to accept high and/or poorly evaluated risks. It is also a fact that something very important is at stake: since it involves the birth of human children, can people be allowed anything they want with no control? However, it isn't clear what we should be trying to achieve. Should we ban cloning as a method of reproduction, and not bother about the rest? This would mean permitting, in the context of MAP, for instance, that deaf children could be deliberately produced on demand, as has in fact happened. Should we compile a list of the genetic interventions that are authorized or banned in MAP (with the risk of seeing these lists rapidly becoming outdated)? Should we, as in France, be tolerant with regard to treatments for sterility, but ban outright all research concerning the technique of nuclear transfer, even though such research is going on elsewhere (in the United Kingdom, Israel, Singapore, for instance)? It is already complicated enough for a single country to adopt an overall policy for reprogenetics (Canada has had problems), it can only be even more difficult at the international level. In 2003 we saw how the efforts of conservative groups to have the UN ban all forms of human cloning failed, and that just a few months later a team of biologists in south-Korea announced that they had produced a human embryo by cloning. Even though that result was proved fabricated, other research groups are now on the same path.

A few philosophers have tried to formulate the principles for regulating technological progress. Jürgen Habermas (2001) finds it quite unacceptable that a human being could, to no matter how limited an extent, become a technological construct (because, for instance, a genetic defect had been corrected). Following Kant's distinction between the realm of nature and that of freedom, and fearful that increasing technicization of human nature could obscure the sense that we have of our dignity, he firmly resists the prospect of an invasion of the field of human procreation by genetic engineering: "If we get the habit of turning to biotechnology to rearrange human nature to fit in with our preferences, it is impossible that the understanding we have of ourselves from the point of view of an ethics of the human species could remain intact" (2002, 109). Claude Debru, tracing the history of the biotechnologies, observes that although we have had fantasies of Frankenstein, in reality we have produced insulin to treat people with diabetes. Having described the position of Habermas, he rejects its rigidity, on the grounds that our biotechnological tinkering is in evolutionary continuity with that which occurs in nature: "the tools and bases of biotechnological tinkering are the same as those that underlie evolutionary tinkering" (2003, 420). He is inclined to prefer a flexible control of biotechnological progress, an approach which would keep confidence in the ability for optimization inherent in human nature, and which would avoid compromising the future: "we must be careful not to go into the future facing backwards" (ibid., 409).

5. CONCLUSION

To conclude, let us return to the clinicians' view, which is less tragic or all-encompassing than the anthropological concerns that we have been discussing. Worried that the medical applications of genomics could further increase the gulf between the countries of the north and those of the south, in January 2003, two scientists (Alwan and Modell, 2003) published a proposal in *Nature Genetics*: they argued that some genetic diagnostic tests are now so easy and cheap that they could be made as readily available as measuring someone's temperature, blood pressure, or blood glucose; and that where health services exist at all, they could and should include these tests in what the UN has described as primary care to be provided in dispensaries. In addition, in a journal that makes a considerable contribution to the continuing GP education in France (see in *Concours Médical* 2003 the interview with Gilles Vassal) a researcher from the Institut Gustave Roussy (IGR, cancer institute) highlights the importance of two learned articles that were published recently (see Van de Vijver et al., 2002; Ramanawamy, 2003), showing how within the next few years, genomics could begin to impact on oncological practice. The problem is that of preventing metastases in the context of breast cancer. Up to 90% of women who have simple breast cancer (*i.e.*not involving the lymph nodes) receive adjuvant chemotherapy in addition to the standard treatment (such as surgery), and it would seem that half of them do not need it. What the article in *Nature Genetics* does is to demonstrate the link between the occurrence of metastases and a particular genetic 'signature'. The article in the *New England Journal of Medicine* reports the findings of a retrospective study, carried out in a population of women who have had a "node negative" breast cancer, and of whom we know whether they went on to develop metastases. Using a tool known as a 'DNA chip', the expression profile of the genes in these women was determined, and this made it possible to identify a set of 70 genes that were expressed differently depending on whether the cancer did or did not go on to develop metastases. To assist GPs, the researcher explains clearly that these results alone do not suffice. There still needs to be a prospective study, in order to find out whether this genetic "signature" has sufficient predictive value to make it possible to adapt the treatment strategy in the light of the results of the test. Such a study will be expensive, and take several years to perfect: no change in treatment is therefore going to happen tomorrow, although the Company that manufactures the test has gone ahead to patent and market it without delay. The same approach has begun to be adopted to other cancers, and small signs like these give us a glimpse of how genetic technologies can be introduced into clinical practice, for the benefit of patients, and without leading to any major anthropological collapse.

REFERENCES

Abir-am Pnina G (2004) DNA at 50: Institutional and biographical perspectives. Minerva 42:191–213.

Alwan A, Modell B (2003) Recommendations for introducing genetic services in developing countries. Nat Rev Genet 4 Jan, 61.

American Society for Reproductive Medicine (2000) Does ICSI carry inherent genetic risks? ASRM, Birmingham, Alabama.

Barritt JA et al (2000) Spontaneous and artificial changes in human ooplasmic mitochondria. Hum Reprod 15 (Special suppl 2):207–217.

Barritt JA et al (2001) Mitochondria in human offspring derived from ooplasmic transplantation. Hum Reprod 16(3):513–516.

Bell JI (2003) The double helix in medical practice. Nature 23 Jan, 421:414–416.

Cohen J et al (1998) Ooplastic transfer in mature human oocytes. Mol Hum Reprod 4:269–280.

Concours médical (2003) Genetic signatures used for therapy. The identification of expression profiles of genes associated with the prognosis should make it possible, in the future, to adapt the management to the risk facing each patient in the field of oncology. September, 125–25, 1422–1424.

Condit C, Williams M (1997) Audience responses to the discourse of medical genetics: evidence against the critique of medicalization. Health Commun 9 (3):219–235.

Debru C (2003) –avec la collaboration de Pascal Nouvel– Le possible et les biotechnologies PUF, Paris.

Delay J, Deniker P (1952) Trente-huit cas de psychoses traitées par la cure prolongée et continue de 4560 RP. Congrès de psychiatrie et de neurologie de langue française Masson, Paris.

Feldman M, Tauber A (1997) Sickle cell anemia: reexamining the first molecular disease. Bull Hist Med 71(4):623–650.

Fugger EF et al (1998) Birth of normal daughters after microsort sperm separation and intrauterine insemination, in vitro fertilization, or intracytoplasmic sperm injection. Hum Reprod 13:2367–2370.

Gordon JW (1999) Genetic enhancement in humans. Science 283:2023–2024.

Habermas J (2001) Die Zukunft der menschlichen Natur. Auf dem Weg zu einer liberalen Eugenik? Suhrkamp, Frankfurt.

Hansen M et al (2002) The risk of major birth defects after intracytoplasmic sperm injection and in vitro fertilization. N Engl J Med 346:725–730.

Hedgecoe A (1998) Geneticization, medicalisation and polemics. Med Healthc Philos 1(3):235–243.

Hedgecoe AM (2001) Ethical boundary work: geneticization, philosophy and the social sciences. Med Healthc Philos 4(3):305–309.

Hoedemaekers R, ten Have H (1998) Geneticization: the Cyprus paradigm. J Med Philos 23(4):274–287.

Holtzman NA, Marteau TM (2000) Will genetics revolutionize medicine? N Engl J Med 343:141–144.

Husserl E (1954) Die Krisis der europäischen Wissenschaften und die transzendentale Phänomenologie (1935–1936). Martinus Nijhoff, The Hague.

Illich I (1977) Limits to medicine. Medical nemesis: the expropriation of health. Harmondsworth (Penguin), New York.

International Molecular genetic Study of Autism Consortium (1998) A full genome screen for autism with evidence for linkage to a region on chromosome 7q. Hum Mol genet 7(3):571–578.

Stéphane J et al (2003) Mutations of the X-linked genes encodage neuroligins NLGN3 and NLGN4 are associated with autism. Nat Genet, May <www.*nature.com/ng/*>.

Laborit H, Huguenard P, Alluaume R (1952) Un nouveau stabilisateur végétatif, le 4560 RP. Presse médicale 60:206–208.

Leboyer M, Bellivier F (2003) Psychiatric genetics. Methods and Reviews. Humana Press, Totowa, NJ.

Lippman A (1991) Prenatal genetic testing and screening: constructing needs and reinforcing inequalities. Am J Law Med 17(1–2):15–50.

Le BD (1993) La chair à vif. Usages médicaux et mondains du corps humain. Métaillé, Paris.

Morange M (1998) La part des genes. Odile Jacob, Paris.

Nelkin D, Lindee MS (1995) The DNA mystique: the gene as a cultural icon. Freeman, New York.

Pauling L, Itano HA, Singer SJ, Wells IC (1949) Sickle cell anemia, a molecular disease. Science 110:543–548.

Parens E, Knowles LP (2003) Reprogenetics and public policy. Reflections and recommendations. A Special Supplement to the Hastings Center Report Jul–Aug 33(4):S1–S24.

Ramanawamy S et al (2003) A molecular signature of metastasis in primary solid tumors. Nat Genet 36:38–44.

Rostand J (1950) Uchronie scientifique. La biologie et l'avenir humain, Albin Michel, Paris.

Templeton A (2002) Ooplasmic transfer: proceed with care. N Engl J Med 346:773–775.

ten Have HAMJ (2001) Genetics and culture: the geneticization thesis. Med Healthc Philos 4(3):295–304.

Van de Vijver M et al (2002) A gene expression signature as a predictor of survival in breast cancer. N Engl J Med 347:1999–2009.

Verlinsky Y et al (2001) Preimplantation diagnosis for Fanconi anemia combined with HLC matching. JAMA 285:3130–3133.

World Federation of Neurology: Research Committee Research Group on Huntington's Chorea (1989) Ethical issues policy statement on Huntington's disease molecular genetics predictive test. J Neurol Sci 94:327–332.

World Health Organization (2001) The World Health Report 2001. Mental health: new understanding, new hop.e Geneva: OMS <www.who.*int*/w*hr*2001>.

CHAPTER 10

THE HARM OF BEING A CLONE

JEAN-YVES GOFFI

Université Pierre Mendès France, Grenoble 2, UFR SH, Grenoble, France
Jean-Yves.Goffi@upmf-grenoble.fr

Abstract: It has been said that, even if human reproductive cloning become some day a safe and secure procedure, a clone would suffer some sort of harm. This assertion obviously depends on the way the concept of harm is defined. It is argued that someone is harmed when living in a condition that is harmful, no element of comparison with a previous unharmed condition being required. It is concluded that a clone would not suffer any ontological or existential harm; but it would certainly suffer some psychological distress, resulting from irrational expectations on it

In the early 90s, K. Bayertz predicted "the opening up of a whole new continent of possible human actions" (Bayertz, 1994, 1). The starting point of this alleged revolution was the birth, on July 25th, 1978 of Louise Brown, the first 'test-tube baby' in human history. According to K. Bayertz, this first birth following IVF meant that we were on the threshold of what S. Lem had called the second stage in autoevolution, that is, the gradual creation over hundreds or maybe thousands of years of 'the next *homo sapiens* model'. According to the proponents of this model (such as J. Lederberg and S. Lem himself), human being was no longer considered as a macroscopic organism, but as a biochemical machine whose constituents could be modified depending on options afforded by technology. In a nutshell, the meaning of the birth of Louise Brown was assessed as follows: "Evolution is replaced by construction" (Bayertz, 1994, 75). In his book, K. Bayertz does not lay great stress on human cloning, but one can easily understand why this practice fits well into the radical constructivist perspective supported by S. Lem. As early as 1966, J. Lederberg asserted:

> If a superior individual – and presumably, then, genotype – is identified, why not copy it directly, rather then suffer all the risks, including those of sex determination, involved in the disruption of recombination?"
>
> (Lederberg, quoted in Bayertz, 1994, 69)

A. Fagot-Largeault et al. (eds.), The Influence of Genetics on Contemporary Thinking, 151–163.

Of course, by using the word 'superior', J. Lederberg commits a blunder, since the term is only too reminiscent of old-fashioned eugenics. But replacing it with 'desired' or even 'planned' still captures the main idea well: by cloning, we can propagate "a genotype already tested in one generation for further trial in a second" (Lederberg, quoted in Bayertz, 1994, 69).

Things really wished for always happen eventually, so that, on July 6th, 1997 a sheep named Dolly was born. Dolly was obtained from a differentiated somatic cell fused with an egg from which the nucleus had been removed. The 'father' of this animal was Ian Wilmut of the Roslin Institute in Scotland. Ian Wilmut and his colleagues seemed to have discovered a method, or at least the means, for making a differentiated adult cell go back to a stem cell state, so that when it was fused with an enucleated egg it behaved like an undifferentiated cell. The trick was that when the donor cell was fused in the membrane of the enucleated egg, adult cell chromosomal DNA was released: the result was a perfect case of asexual reproduction. The cloning of Dolly was thus very different from other procedures known as 'cloning', a very ambiguous term as G. Pence emphasized (Pence, 1998, IX). G. Pence notes that cloning may refer to molecular cloning, in which strings of DNA are replicated in a host bacterium; to cellular cloning in which copies of a cell are made, resulting in a 'cell line'; to embryo cloning, in which an already formed by sexual reproduction embryo is split into two identical halves; and to somatic cell nuclear transfer (SCNT). This procedure can work either by taking the nucleus of an adult cell and implanting it in an enucleated egg, or by fusing donor cells with an enucleated egg: the latter method seems to actually have been implemented by Ian Wilmut.

Once Dolly's birth was made public, it caused a big sensation: was this the first step towards human reproductive cloning? After all, the difference is not so significant, from a conceptual point of view, between producing a lamb and producing a child with SCNT. In the United States, President Clinton asked a commission to investigate the ethics of human cloning. The fifteen members of the National Bioethics Advisory Commission were given only three months to investigate the ethical and legal issues surrounding the subject of human cloning and make recommendations. The latter point was important, as President Clinton had previously introduced a ban on all federal funding of attempts to create a child by means of SCNT. As it turned out, the Commission recommended that this moratorium on the use of Federal funding be continued and that "all firms, clinicians, investigators and professional societies in the private and non-federally funded sectors ... comply voluntarily with the intent of the federal moratorium" (National Bioethics Advisory Commission in Pence, 1998, 61). The method used to reach these conclusions included hearings with representatives from a variety of philosophical and – at the request of President Clinton – religious traditions, and the commissioning of a paper by D. Brock (Brown University) entitled "Cloning Human Beings: an assessment of the Ethical Issues pro and con" (Brock in Nussbaum and Sunstein, 1998). The part of the report specifically dealing with ethical considerations was a carefully worked out assessment of pro and con arguments. The main con arguments were:

1 Potential for physical harms.
2 Potential for psychological harms.
3 Potential harms to the family.
4 Potential harms to important social values.
5 Fear of treating people as objects either by lacking respect toward persons or by coming to consider them as commodities to be exchanged, bought or sold in the market place.
6 Eugenic concerns.

The main pro arguments were:

1 General presumption in favor of personal autonomy.
2 Freedom of reproductive choice.
3 Freedom of scientific inquiry.
4 Special circumstances which would make the choice to create a child by means of SCNT understandable and even desirable.
5 The speculative and unproven nature of many con arguments.

In fact, these arguments underscore the risks and benefits attached to SCNT and it seems clear that the Commission concluded that risks outweighed benefits.

In a way, the Commission adopted a middle of the road solution: the report recommended that federal legislation be enacted to prohibit the creation of a child through SCNT; but this legislation was to include a sunset clause to ensure that the issue was reviewed after a specified time period; and it made it clear that this legislation should be written so as not to interfere with important areas of scientific research: in particular, no new regulations were required which would prohibit the cloning of human DNA sequences or cell lines. It also allowed the principle of cloning animals by means of SCNT, provided it was subject to already existing regulations regarding the humane use of animals in the field of research.

Such moderate conclusions were doomed to be criticized both by opponents of reproductive cloning, who considered all forms of cloning intrinsically wrong, and by supporters of SCNT, who saw this procedure as just one more reproductive technology among many others. D. Callahan's stance was singular in that he did not enlist in either of the opposing camps. Instead, he assessed the debate itself as a symptom of the failure of bioethics to respond to novel biological developments. To quote Callahan:

> Cloning was first debated in the early to mid-1970, and the latest Dolly-inspired round has added little new to what was said at that time.
>
> (Callahan, 1997, 18)

According to D. Callahan, the main voices in the early cloning debate were those of the protestant theologian P. Ramsey, the philosopher H.Jonas, the biochemist L.R. Kass and the jurist F.C. Pizzulli.

Consider, for example, what P. Ramsey published in the early 70s (actually it was written in the late 60s). He focused on proposals made by supporters of neo-eugenics (such as H.J. Muller), on science-fiction scenarios describing the future – and hoped for – reconstruction of man (such as the one written down by

G. Feinberg), and on the writings of some theologians (such as J. Fletcher and K. Rahner) whom he called "techno-theologians". According to P. Ramsey, they all dreamed of complete or at least increased genetic control of man, and they all were blind to the fact that, in their hope of reconstructing humankind, what was really at stake was the humanity of man. At the very least, they were blind to the fact that the 'debiologized procreation' (significantly, they did not talk any more of 'procreation', but were using the manufacturing term "reproduction" instead) which they enthusiastically supported implied a drastic change of meaning for the concept of_human parenthood. As a protestant theologian, P. Ramsey saw sexual intercourse as a unitive – or communicative – good as well as a procreative good. Human sexual relations tend, of their own nature, to strengthen the bonds of love between a man and a woman and to engender a child. As he sees it:

> We procreate new beings like ourselves in the midst of our love for one another, and in this, there is a trace of the original mystery by which God created the world because of His love.
>
> (Ramsey, 1970, 38)

To separate the unitive and the procreative dimensions of sexual intercourse "means a refusal of the image of God's creation in our own" (Ramsey, 1970, 39). Of course, one cannot expect such theological convictions to be universally shared and P. Ramsey was able to articulate them in another register: he deemed the utopian dreams of his opponents to be a well-meaning, but all the more dangerous, attempt to raise human beings above their own condition. But the general refabrication of individuals, along with control of the future of man through genetic manipulation and an alteration of the nature and meaning of human parenthood, are likely to bring out such radical changes in humankind that they "can only be described as the death of the species and its replacement by a species of life deemed more desirable. That I take to be similar to the inner motive and action of any suicide." (Ramsey, 1970, 152). The mistake common to all his opponents is to believe that there can be ethics without ultimates; or better, that human will and thought are such ultimates, without paying attention to the fact that human beings are embodied persons: "an individual's body, including his sexual nature, belongs to him, to his *humanum*, his personhood and self-identity, in such a way that the bodily life cannot be reduced to the class of animal over which Adam was given unlimited dominion. To suppose so is bound to prove antihuman – sooner than later." (Ramsey, 1970, 87). P. Ramsey concludes that fabricating better people or trying to manufacture a better humankind is a perfectly misguided use of medicine:

> Actions whose objective is treatment and actions whose objective is the control of the future of our species are different sorts of actions, even when descriptively they may look alike.
>
> (Ramsey, 1970, 121)

To begin with, it is a perversion of medicine that concern for the species replaces the curing of primary patients. But which is more, it is quite uncertain that people unable to beget children are primary patients: no one has an unqualified right to have children and children are not simply for one's own fruition. If the only way to have a child implies negating the human condition, and such a process occurs each time procreation is debiologized, then the procedure is inhuman and unethical: a thing that should never be done. Of course, human cloning by SCNT ranks high among these acts which can be labelled as intrinsically bad.

We can state the difference between P. Ramsey's and H. Jonas' approaches as follows: P. Ramsey thinks that one of the main wrongs of cloning is its bringing into existence of a human being by way of asexual reproduction; H. Jonas thinks that the main wrong of cloning is its bringing into existence_of a human being who is the exact copy of somebody else. In others words, H. Jonas addresses the problem from the 'subjective' side: he says something like 'what is it like to be a clone?'. He sees clones essentially as twins separated by a time_lag, which enables him to claim, first, that the clone knows (or believes he knows) too much about himself, and second, that the clone is known (or is believed to be known) too much by others (Jonas, 1973, 161). The clone falls prey to a genuine form of tyranny because 'it is the known donor archetype that will dictate all expectations, predictions, hopes and fears, goal settings, comparisons, standards of success and failure, of fulfilment and disappointment, for all 'in the know" – clone and witness alike' (Jonas, 1973, 161). The strength of H. Jonas' approach is that, even if it false as a matter of fact that replication of the genotype entails repetition of an individual's achievements, this will turn out to be irrelevant because the donor has been chosen on the basis of such a belief: the clone will be expected to behave exactly like his donor, even if this expectation is groundless. The argument can be put in another way: the question: 'who am I?' cannot really be asked by a clone. That is because such a question presupposes that one is in fact able to become who he is; and this can be only accomplished by one who genuinely ignores what the result of the process will be. Although H. Jonas does not express his idea explicitly so, we can reconstruct his argument as a 'destructive dilemma':

– either the clone will be as good as his donor, in which case he will have become someone else. – or the clone will be inferior to his donor, in which case his life will be a failure.

Neither alternative is acceptable. H. Jonas notoriously concludes by saying that cloning would violate a right to ignorance by failing to "*respect the right of each human life to find its own way and be a surprise to itself*" (Jonas, 1973, 163).

Can it really be claimed that little has been added to the discussion since? It has to be admitted that the debate is no longer managed in the same way. The paper by D. Brock, for one, is quite dispassionate and does not manifest the sense of urgency one can find in the pioneering works of P. Ramsey and H. Jonas. In a way, that can be explained by the circumstances. D. Brock was asked to provide a survey of pro and con arguments, not to take sides in the debate. The commission was not expected to make important advances in the field of substantive ethics, but rather to reach a consensus, that is, to achieve a result in the field of procedural ethics. But that also means that something like a division of labor occurred as the debate matured – and, perhaps, that it became more boring: the pioneering papers by P. Ramsey and H. Jonas were very creative, but also manifested something like amateurism. H. Jonas, for one, speaks of a right to ignorance. And he correctly observes that such a right has never been rated high, nor claimed for, as lack of knowledge has always been deplored as a deficiency in the human condition and an obstacle on the path of virtue. So, a genuine right to ignorance would really be something new. But, as H. Jonas himself insists, it would be new to "ethical theory" (Jonas, 1973, 161). If I understand him correctly, this means that he is not speaking of a legal right, and this is fortunate because what is at stake, according to H. Jonas, is that the spontaneity of becoming oneself is jeopardized in the case of clones. It seems difficult for any legal system whatsoever to recognize as a right the "spontaneity to become oneself". But a legal system can recognize a right to freedom of action, and many have actually recognized such a right. From such a perspective, if a clone was harmed in a 'jonasian' way it would be harmed because it was born in a condition such that no open options in respect of some possible course of action would ever exist for it: something in its objective circumstances would always prevent it from doing X if it should so choose; and something in its objective circumstances would always require it to do X if it should choose not to (Feinberg, 1986, 64). But of course, that has nothing to do with a right to ignorance and J. Feinberg himself, when taking this approach, talked of a right to an open future (Feinberg, 1992). Such confusion between legal and moral rights does not speak in favor of H. Jonas's analysis. As far as this kind of confusion does not appear to be made any more, it looks as if the debate had improved since the early 1970s. Another difficulty for D. Callahan is the following: his standpoint seems to imply that he would not attribute much importance to the fact that a current major pro argument in the cloning debate does not seem to have been appealed to much in the 1970s: the right to reproductive freedom or procreative liberty. Here, he seems to overlook powerful arguments such as those advanced by J.A. Roberston (Robertson, 1998) a Professor of Law at the University of Texas, who ranks among supporters of reproductive human cloning. Robertson argues that procreative liberty – or reproductive freedom – i.e., the freedom to decide whether or not to have offspring, can also be described as a moral right to reproduce. Then he says that the "moral right to reproduce does include the right to use noncoital or assisted means of reproduction" (Robertson, 1998, 1390). J.A. Robertson constructs SCNT as an assisted mean of reproduction; he can then assess, taking the uncontroversial case of a married (heterosexual) couple with an infertility problem, whether the possible harms associated with the implementation of these means may justify any restriction or prohibition. He concludes that serious cases for restriction or prohibition are much less widespread than one might initially believe. Nor does D. Callahan seem very interested in the emergence of new voices in the debate: gay and lesbian groups and, more generally, people who talk of 'families of choice'. Their arguments may not be very sound from a logical point of view – and they do not seem very sound indeed to this author; but the very fact that they are asserted does seem to add something new.

In the end, it looks as if the genuine main thread of the cloning debate has been_the problem of harm caused. The question H. Jonas already set himself was:

'Is one harmed in being someone's else clone?'. And his answer, as has been recalled, was affirmative. A clone will, of necessity, suffer a major harm, a kind of ontological or, as H.Jonas says, existential harm: 'The trial of life has been cheated of its enticing (also frightening) openness' (Jonas, 1973, 162). And the National Bioethics Advisory Commission stresses that virtually all people agree that there exist major risks of physical harm to the children cloned. But can harms other than physical – for example 'jonasian' harms – be also taken into account? The Commission, while being consensual as to the current unsafeness of the procedure, is less affirmative when mention is made of harms other than physical: suddenly, it is not a question of lack of safety, only one of concern relating to potential psychological harms whose speculative nature may justify prohibition in the future, especially if compounded with effects on important moral, religious and cultural values of society. But giving content to the concept of non-physical harm is harder than one would believe at first glance.

It has been maintained that a clone cannot suffer harm, except if born in a very painful and hopeless condition. D. Brock takes R.Chadwick and J. A. Robertson to have supported such a claim. But in fact their conclusions are not identical. R. Chadwick wants to argue that in the cloning case no right to genetic uniqueness is violated: the person who is born as a result of cloning cannot be said to have the right to be genetically unique 'because *he* would not have existed if he had not be cloned' (Chadwick, 1982, 204). In order to understand the point of this argument, let us examine a misinterpretation of it. On this misinterpretation, the argument would run as follows: for a right to be violated, one must have

1) A right bearer existing at time t.
2) An act which is a violation of the right bearer's right, occurring at time t1 (t1 being posterior to t).

But in the cloning case, there exists no right bearer at a time t anterior to the violation of his alleged right, occurring in t.1) The bringing of the right bearer_into existence occurs exactly at the very same moment that his alleged 'right to genetic uniqueness' is violated. So condition 2) is not fulfilled and no right is violated. But that will not do. Let us suppose that there exists a right to a safe and clean environment; then, let us suppose that a person is conceived at a time and place where the_environment is dangerous and unclean, with no probability that things have improved by the time she is born. Whether one equates 'bringing into existence' with conception or with birth, or with some event between conception and birth, is irrelevant: the right to a safe and clean environment is violated at the very same moment the person is brought into existence. And yet no one would take this to be a proof that there cannot be such a thing as a right to a safe and clean environment. So it seems that what is relevant here is not the time lag between the bringing into existence of the right bearer and the violation of his right. R. Chadwick probably wants to say that no right can be formulated in such a way that the voluntary act which brings the right bearer into existence is numerically the same as the voluntary act which violates his right. In the former counter-example, the right to a safe and clean environment would have been contingently violated in that it is not the same act which brings the person into existence and makes the environment unsafe and dangerous, thus violating his right. But if a right to genetic uniqueness is recognized, we shall have a class of acts which, of necessity, will bring into existence right bearers and, at the same time, will violate this right: acts of cloning.

But it is one thing to see one's rights violated and quite another to be subjected to some harm. If an earthquake destroys my house, I certainly suffer some harm; it does not follow that a right of mine has been violated.

This distinction is clearly made by J.A. Robertson. In a 1994 paper, he attempted to show the ethical acceptability of splitting off cells from embryos in order to produce cloned human beings. One of his arguments is that the procedure cannot wrong the children thus produced. He goes as far as saying that a multifetal pregnancy resulting from such a procedure will not put the children born prematurely in a wrongful life situation even if 'they end up with permanent learning and physical disabilities' (Robertson, 1994, 10). J.A. Robertson is not very interested in the niceties lying behind the wrongful life problem as he considers his own position to be equivalent, from a practical point of view, to the rather different one held by D. Heyd. J.A. Robertson argues that: 'preventing existence as a way to prevent harm to the person who would exist makes sense for that person only if it reasonably appears that once born, the child existence would be so full of pain and suffering that its interests would be best served by non-existence' (Robertson, 1998, 1405). On the other hand, D. Heyd argues that someone cannot be benefited or harmed by simply being born; but once born, someone has an interest in the continuation or cessation of one's own life. The key difference is that J.A. Robertson seems to say that one can be benefited or burdened by being born, whereas D. Heyd would see birth as neutral as to whether interests are enhanced or damaged; nevertheless, the conclusions of the two arguments are quite similar, as the condition which would militate in favor of cessation of the newborn's life is hard to distinguish from the condition in which a newborn would see his interests best served by non-existence. However that may be, J.A. Robertson asserts that 'genetic defects, such as sickle cell anemia, cystic fibrosis, or Down's Syndrome, do not have such devastating effects that a child born with those conditions would be better off, from its own perspective, never living at all.' (Robertson, 1998, 1405). Clearly, this clause is meant to avoid the conclusion that the life of someone with a genetic – or other – defect, however small, is not worth living. As a consequence, J.A. Robertson considers that only a condition such as that of a child born with Tay Sachs disease – or any other similar disease – would be a wrongful life condition, because of the harm and suffering involved. He can then easily conclude that human cloning, if shown to be safe and effective, is not likely to bring about such a condition. This amounts to saying that, for a life to be wrongful, the 'price to be paid' must be very high indeed.

But is it really necessary to refer to the wrongful life question in order to assess the harm of being a clone? One can notice that J.A. Robertson's analysis of the wrongful life problem is, in one crucial respect, quite similar to J. Feinberg's, who wrote what is probably the major recent contribution to the topic. J. Feinberg clearly distinguishes between the concept of a harm and the concept of a wrong because he wants to stay in line with the conceptual tradition of the Law of Torts: 'to be harmed is to be put in a worse condition than one would otherwise be in (to be made "worse off")' (Feinberg, 1984, 102). So a child born with a terrible disease which makes his life not worth living can very well be said to be born in a harmful condition; but he cannot be said to have been harmed: there is no prior condition of his which would be his unharmed condition. But he can be said to have been wronged, because he 'comes into existence with his more basic "birth rights" already violated' (Feinberg, 1984, 102). Here we see the right to an open future in action as we have a situation preventing the unfortunate newborn from making autonomous choices; and we understand why the disease must be terrible indeed: some conditions such as achondroplasia or congenital deafness still allow autonomous choices. To be in such a condition does not mean that 'birth rights"' are denied. But when someone suffers horribly and can make no autonomous choices whatever, then such rights are violated. The newborn is wronged because non-existence is preferable to his present condition, although he is not harmed as he has not been made 'worse off'. There is a difficulty here, as neither J. Feinberg nor J.A. Robertson want to say that there exists an objective standard which, correctly applied, could allow the conclusion that a life is not worth living. J. Feinberg imagines a kind of debate between negligent parents and the newborn, the latter arguing that non-existence was the preferred alternative. And J.A. Robertson notices that a child affected with sickle cell anaemia or Down's syndrome can judge, from his own perspective that his own life is not, after all, such that he would be better off never living at all. The point is that for a newborn to be better off never living at all, his condition must be so terrible that he is likely to lack any perspective of preference. In fact, J. Feinberg seems to accept such a conclusion by making the unfortunate newborn say to his negligent genitors: 'Any rational being ... would prefer not to exist than to exist in such a state' (Feinberg, 1984, 102).

But let us put this difficulty aside. J. Harris takes a different approach to the question in saying that 'to be harmed is to be put in a condition that is harmful' (Harris, 1993, 88): no element of comparison is required here before deciding that harm has occurred. And there is no need to hesitate between an internal and an external perspective because a harmed condition is 'one in which the individual is disabled or suffering in some way' (Harris, 1993, 89). This suggests a quite different analysis of the relations between harming and wronging: to be harmed is to be in a condition in which one is disabled or suffering in some sense; to be wronged is to be deprived of a net benefit and to have some rights violated. Someone can be harmed without being wronged, as would be the case of a victim of a plane crash whose life is saved by emergency surgery and who survives with moderate permanent after-effects.

So, can it be said that we should not clone people because they would be harmed – put in a disabled or suffering condition – in a 'jonasian' way? We have seen that H. Jonas does not think that a clone would be harmed because of being denied some right to uniqueness. And H. Jonas is penetrating enough to put aside a right to uniqueness of genotype and considers that what really matters is a right to 'uniqueness of being' (Jonas, 1973, 160) upon which the famous 'right to ignorance' supervenes. But is such a right violated by SCNT? It should be pointed out that there is some confusion about what is meant by uniqueness. For example, the NBAC asks: 'Is there a moral or human right to a unique identity, and if so, would it be violated by this manner of human cloning?' (National Bioethics Advisory Commission in Pence, 1998, 50). But the Commission's position is unclear because such terms as 'distinct', 'unique', 'autonomous' are taken to be more or less synonymous terms referring to what, in the end, turns out to be the condition of an individual enjoying "a unique qualitative identity" (National Bioethics Advisory Commission in Pence, 1998, 50). But what must we understand by 'qualitative identity'? Qualitative identity is usually contrasted with numerical identity: two things are numerically identical if they are one and the same; two things are qualitatively identical – or indiscernible – if they instantiate the same qualities. K. Evers (Evers, 1999), in what is the most convincing attempt to take seriously the claim that clones may be identical, has argued that what worries people about SCNT is not that clones could be numerically identical; the real worry is that they could be qualitatively identical, that is, indiscernible as to their qualities. This cannot amount to indiscernibility in respect of relational properties, since clones are spatio-temporally distinct. So it must be indiscernibility relative to intrinsic properties, those that a clone would possess 'independently of other things' (Evers, 1999, 69), the qualities the clone would retain in case it was the only being in the universe. Lastly, those properties must be those that a clone would possess as a matter of fact, those which constitute 'empirical identity' or individuation. It is then possible to show that a clone has, in itself, empirical properties that individuate it from any other organism: it manifests genetic, physiological, perceptual and cognitive individuation and has a personality of its own. So a clone is a perfectly respectable entity as to its own individuation: there is at least a difference at the genetic level as there is a unique contribution of mithocondrial DNA from the enucleated egg. Of course, at 'higher' levels, more differences can be spotted. And one is probably entitled to argue that from an internal perspective, these differences have much more weight than mere genetic differences. Only someone who had the crudest conception of genetic determinism could believe that a clone and his donor are not individuated, in the sense that they share all their empirical properties. Furthermore, such a belief is self defeating because someone who admits this crude conception of genetic determinism believes that two people will be exactly alike if and only if they have exactly the same genetic material; but, as a matter of fact, a clone and its donor *do not have* exactly the same genetic material. As for genetic individuation, a clone and its donor are certainly less indiscernible than natural identical twins. So, strictly speaking, it does not make much sense to assert that a clone may be harmed because it lacks individuation.

But that does not fully settle the matter as to the inflicting a 'jonasian' harm to a clone. For H. Jonas does not intend to argue for a right to uniqueness of being, nor for a right to uniqueness of genotype. According to H. Jonas a clone is harmed, in the first place, because its biological life has been denied its openness; and because, in the second place, the question 'who am I?' has been answered before it could be fully articulated.

As for the first type of harm, it depends on H. Jonas' 'philosophical biology', according to which life is 'viewed as an experiment with mounting stakes and risks which in the fateful freedom of man may end in disaster as well as in success' (Jonas, 1966, X). But much more rests on this philosophical biology than the definition of a type of harm. H. Jonas thinks that in living creatures the natural purpose of organic life, that is self-affirmation 'becomes increasingly subjective, that is, increasingly the individual executants' very own' (Jonas, 1984, 81). This is what is denied to a clone as far as its production is concerned: here "the past has been made to preempt the future" (Jonas, 1973, 162). But why should it be so? Suprisingly, H. Jonas does not lay great stress upon the fact that a clone is fabricated and so reduced to an inorganic mode of being brought into existence. He rather thinks that what really matters is the answer to the question: 'Why should cloning be made at all?'. According to him, the answer is as follows:

> A known individual life performance sufficiently outstanding in some desired respect to prompt the wish for more of its kind, and rare enough in its (presumed) genetic basis not to expect the desired frequency of it from the chances of ordinary or even selective interbreeding.
>
> (Jonas, 1973, 156)

This line of reasoning seems open to criticism. First of all, it is not obvious that there is only one motive for resorting to SCNT and that that is precisely the one H. Jonas relies on. In the second place, H. Jonas' philosophical biology is highly speculative. This is not to deny its interest, but can it be asserted as a matter of fact that in the purposiveness in nature 'we can see a fundamental self-affirmation of being, which posits it absolutely as the better over against nothingness' (Jonas, 1984, 81). But if we cannot resort to such a philosophical biology, what are we left with? We certainly cannot say any more that in cloning the trial of life will be cheated of its ambiguous openness. But we can still say that a clone can be harmed because of irrational expectations that people wanting to resort to SCNT will have. And it makes sense to say that a clone's identity may be jeopardized, not because the clone will be identical to Einstein or Mozart, but because people will expect that it will be another Einstein or Mozart. In this case, of course, we do not speak about logical identity, not even maybe about individuation. We rather speak of something like a sense of identity which correlates with the development of the individual's personality. This sense of identity may be roughly defined as individuation from an internal perspective. And one can certainly be harmed, that is put in a harmful condition, if other people's irrational expectations blur this internal perspective. Such a condition would be one of psychological distress, especially if one considers that a clone is brought into existence by being 'manufactured' in a way. It is already difficult to live with the idea that one has been conceived in order to fulfil somebody's irrational expectations; it may be still more difficult to live with the idea that one has been produced in order to fulfil such expectations. So a clone may be harmed in a 'quasi-jonasian' way.

But assuming the SCNT is safe and effective – which is obviously not the case at the present time – a clone can be the subject of be more or less harm, according to the intentions of people wanting to produce it. Can we sort out those intentions, and on what basis? Of course, intentions of a certain type are at stake here, not particular intentions of particular individuals, as they may remain opaque for ever. It seems that two criteria could be retained as significant. The first would rest on the kantian test: 'Will the clone, once born, be treated solely as a means?'. This rules out use of SCNT to secure a source of organs for example; and renders suspect the use of reproductive cloning to replace a dead or dying child. But of course, there are cases which are not clear cut: it is one thing to try to get bone marrow from a 'younger' twin; it is another to try to get a kidney in such conditions. So the 'kantian' criterion may be supplemented with a further criterion (as suggested by J.A. Robertson, 1998, 1400). Here, one would try to establish the finality of the act of cloning: is it reproduction or replication that is intended? The greater the reproductive intention is, the better chances there are of containing 'quasi-jonasian' harm at a tolerable level. But the greater the replicative intention is, the more likely it is that 'quasi-jonasian' harm will rise above a level deemed tolerable.

ACKNOWLEDGMENTS

I thank my colleague and friend John Zeimbekis for his help in improving the style of this paper.

REFERENCES

Bayertz K (1994) Genethics. Technological intervention in human reproduction as a philosophical problem, Cambridge University Press, Cambridge [1987].

Brock DW (1998) Cloning human beings: an assessment of the ethical issues pro a nd con. In: Nussbaum MC, Sunstein CR (eds) Clones and clones. Facts and fantasies about human cloning. W.W. Norton, New York-London, pp 141–164.

Callahan D (1997) Cloning: The work not done. The Hastings Center Report 27(5):18–20.

Chadwick R (1982) Cloning. Philosophy 57:201–209.

Evers K (1999) The identity of clones. J Med Philos 24(1):67–76.

Feinberg J (1984) Harm to others. The moral limits of the criminal law. Oxford University Press, New York.

Feinberg J (1986) Harm to self. The moral limits of the criminal law. Oxford University Press, New York

Feinberg J (1992) The child's right to an open future. In Freedom and fulfilment. Philosophical essays. Princeton University Press, Princeton [1980], pp 76–97.

Harris J (1993) Wonderwoman and superman. The ethics of human biotechnology. Oxford University Press paperbacks, Oxford, NY [1992].

Jonas H (1966) The phenomenon of life. Toward a philosophical biology. The University of Chicago Press, Chicago & London.

Jonas H (1973) Biological Engineering – A Preview. In Philosophical essays. From ancient creed to technological man. The University of Chicago Press, Chicago and London, pp 141–167.

Jonas H (1984) The imperative of responsibility. In search of an ethics for the technological age, The University of Chicago Press, Chicago & londres, 1984 [1981].

National Bioethics Advisory Commission (1998) Cloning human beings. In: Pence G (ed) Flesh of my flesh. The ethics of cloning humans, Rowman & Littlefield, Lanham, pp 45–65.
Pence G (1998) Introduction in: Pence G (ed), Flesh of my Flesh. The ethics of cloning humans, IX–XV, Rowman & Littlefield, Lanham.
Ramsey P (1970) Fabricated man. The ethics of genetic control, Yale University Press, New Haven and London.
Robertson JA (1994) The question of human cloning. Hastings Center Report 24(2):6–14.
Robertson JA (1998) Liberty, identity, and human cloning. The Tex Law Rev 76(6):1371–1456.

CHAPTER 11

CHILDREN OF ONE'S OWN

Genes, parenthood and the illusion of control

JONATHAN MICHAEL KAPLAN

Philosophy Department, Oregon State University, Corvallis, USA, Jonathan.Kaplan@oregonstate.edu

Abstract: Technologies for assisted reproduction often aim explicitly at giving hitherto infertile couples a 'child of their own' – that is, a child that is genetically related to them. And many couples find themselves spending enormous amounts of money, time, and energy attempting to have a child via these techniques. But why should a genetic relationship make a child any more 'one's own' than other kinds of relationships – for example, those parent-child relationships forged through adoptions? There is a wide-spread assumption in much of contemporary society that genetic parenthood is important because of what it implies about the relationship between the (physical and behavioral) traits of the parents and those traits of the child; arguments relying on these assumptions have even been accepted in some legal cases. I argue here that this state of affairs is particularly unfortunate, and that the over-blown rhetoric of the Human Genome Project and related research programs is at least partially to blame. This rhetoric includes the metaphorical language of genes as 'master controllers', 'blue-prints', 'recipes', and as 'carrying information'. But as none of these metaphors is well-justified by contemporary understandings of the roles played by genes in the organismal development, the metaphors ought to be rejected, and with them, the social emphasis on a genetic relationship as the most important aspect of parenthood

More than anything in this world, I want a child of MY own. When the time is right for my husband and I, we plan to do whatever it takes to make that dream come true.

Anonymous visitor to babycenter.com 12/16/2002,
http://www.babycenter.com/comments/preconception/fertilityproblems/6111

1. INTRODUCTION: A CHILD OF ONE'S OWN

The desire to have a 'child of one's own is often cited as the reason that people are willing to pay extraordinary amounts of money for, and go through enormous discomfort during, various infertility 'treatments', even when they know that the 'treatments' are unlikely to be successful (see The New York State Task Force

A. Fagot-Largeault et al. (eds.), The Influence of Genetics on Contemporary Thinking, 165–180.

on Life and the Law, 1998, 423). Indeed, the most vociferous defenses of human cloning have come from people who believe that the right to have a 'child of one's own' should not be limited in any way, and certainly not by restrictions on the technology used (see Human Cloning Foundation, 1998). But what does it mean to have a 'child of one's own', and why is it considered so desirable?

The first question, at least, seems to have a fairly straightforward answer in contemporary western contexts (perhaps especially in the U.S.). Having a 'child of one's own' in these contexts is taken to mean having a child that is related to you in a particular way, specifically, as your *genetic* offspring. Partially, this relationship involves sharing a particular amount of genetic material – presumably at least 50%. But this relationship is also taken to involve a causal component – that is, the genetic material in question must have come *from you* (have once been a part of you). It is, I take it, for this reason that the son or daughter of your identical twin brother (for example) is not also considered to be *your* son or daughter. And it is equally for this reason that children that result from contract pregnancies where in vitro fertilization was used are generally thought to be the children of the woman whose fertilized egg cell was used, rather than the woman who gestated the children in question (see Bender, 2003).

The second question, why having a 'child of one's own' in the above sense is considered so desirable, is considerably harder to answer. Why, for example, isn't adopting and raising a child generally considered equally desirable? Or, to move the question back to the nature of having a child of one's own – why isn't a child one has adopted and raised just as much a 'child of one's own' as is one that is related by genetic descent?

Part of the answer, I want to suggest, is given by our cultural obsession with genes and genetics more generally. Because of the importance placed on genetics – both in appropriate and, perhaps more so, in inappropriate contexts – the idea that genetic relationships are *the* relationships that count is taken quite seriously, in both legal and more broadly social contexts. Despite claims to the contrary, the rhetoric of genetic determinism and essentialism is alive and well, and is used extensively in defending the decisions of people to spend enormous amounts of money, time, and energy attempting to have 'their own' children through the technologies offered by fertility clinics. The desire to have 'a child of one's own' is in part the desire to have a child that resembles his or her parents in some key ways – the lesser desirability of adoption stems in part from the fear that the adopted child will not resemble the parents in those key ways. What is at the center, then, of the desire to have a child of one's own is the desire to have or impose a kind of control on the features of the child to be parented.

The focus on control makes sense of the current paranoia in the United States about the maternal gestational environment, as well. One often heard worry about adoption is that one might be adopting 'damaged goods' – the mother may have drank alcohol or used drugs during pregnancy, and hence the child might have (currently hidden or cryptic) mental or physical problems that will only emerge later (see Roseman, 2004; Kirby and Hardesty, 1998; see also The New York State Task

Force, 1998, 97) . With one's own child, the idea is, one can control the gestational environment, and hence give the child-to-be the developmental resources necessary for the best chance at an ideal start in life.

But the ability to control the fate of one's child is, at heart, illusionary. Neither 'the right' genes nor 'the right' environment can guarantee a child with 'desirable' features, and furthermore children with less 'desirable' features are usually loved and cherished as much as those with more 'desirable' features. A genetic relationship does not guarantee a child that 'looks like' you, or at least not one that looks very much like you, and there is no evidence to suggest that an adopted child will necessarily suffer the so-called 'genealogical bewilderment' supposed to effect children who are not raised by their genetic parents (Sorosky, 1995). The developmental environment *in utero* is complex and little-understood; while we do know some of the things that can massively derail development, our knowledge of the effects of more minor potential disturbances is still weak at best. Again, to pretend that we know how best to control the gestational environment, and hence that an adopted child is much more likely to have suffered a damaging gestational environment, is to give ourselves far too much credit – it is to posit our having a degree of control that we in fact simply lack.

In what follows, I will argue first that the contemporary social context, at least in the United States (and perhaps in some other broadly Western nations), still continues to favor interpretations of genetic relationships that make genes out to be of fundamental importance to shaping one's features, and hence, to shaping the features of one's children. I will then suggest some reasons why these views should be rejected, and of how a more accurate understanding of the nature of development can undermine views that stress the overwhelming importance of the genetic relationship, without thereby implying that the environment has a similarly overwhelming role. The fundamental lesson here is that the complexities of the developmental process should undermine our faith in *both* the efficacy of genes *and* the efficacy of the environment in shaping, in any controllable way, our children. Finally, I will suggest that once we accept that raising a child is essentially a risky enterprise, the importance we place on the child being *genetically* related to us can be reduced; the risks inherent in adopting a child are not different in kind from the risks inherent in 'normal' procreation, and are perhaps not much different in degree, either. Focusing on the importance of actively embracing the risk of raising *any* child can point towards an interpretation of what it means to have 'a child of one's own' that relies not on genetic relationships, but rather on the social facts of embracing those risks and forging a family.

2. GENES, GENETIC DETERMINISM, AND CHILDREN

It is axiomatic that genetic lineage is important. Parents go to inordinate lengths-physically, emotionally, and financially – to produce genetically related children. Adopted children often invest years in seeking out their genetic parents. Genetic parents search for the children they surrendered for adoption. Children born from sperm donation attempt to contact their biological fathers. Sperm donors seek to establish parental rights to the children fathered through their donor sperm. Genes determine not only physical characteristics, but they also influence personality traits, temperament, preferences, and behavior. In addition, understanding one's genetic heritage and genealogical line is integral to shaping one's identity. Simply put, genes matter.

(Hurwitz, 2000, 149–150)

In writing about the importance of human genetic research, researchers often make claims about the importance of genes to our development and to our personal identities more generally. For example, Gilbert has referred to DNA as 'the most fundamental property of the body', and has claimed that 'there is no more basic or more fundamental information that could be available' then our DNA sequence – it is, he claims, 'what makes us human' (Gilbert, 1992, 83–84). More boldly he states that actually being able to hold one's own gene sequence in one's hand (on CD) will 'be difficult for humans' since 'we look upon ourselves as having an infinite potential' and the recognition 'that we are determined... by a finite collection of information that is knowable' is the 'closing of an intellectual frontier' (Gilbert, 1992, 96). Researchers like Hood refer to the genome as 'our blueprint for life' and the Human Genome Project as creating 'an encyclopedia of life' (Hood, 1992, 136). Mayr notes that DNA 'serves as a blueprint, as a set of instructions' and that this justifies our referring to a 'genetic program' that is "translated into individual organisms" (1982, 827–282). Baltimore claimed that the completion of the sequencing of the human genome provided us with the 'information needed to create a human being' (2001, 814). While Dawkins resists the 'blueprint' imagery, preferring to think of 'genes' as providing a 'recipe' for development (1987, 295–196); he claims that organisms are 'put together' by 'following programs, sets of instructions that organisms carry around inside themselves' and that genes are these programs (1987, 112).

These kinds of claims are made by various human genetics researches in countless interviews and articles meant for popular consumption; such claims have clearly had an impact on the social and legal landscapes surrounding parenthood. For example, in the contract pregnancy ('surrogate mother') case of *Johnson* v.*Calvert*, Johnson (the gestational/birth mother) and Calvert (the woman whose egg was used in IVF and implanted in Johnson) each claimed to be the legal mother of the child that resulted. In this case, both the initial court ruling, and the decision of the appeals court, made explicit reference to the importance of the genetic relationship in framing their decisions. Judge Richard Parslow (a trial judge in Orange County) ruled that the gestational mother in the case was 'analogous to a "foster parent"' and that her womb was merely 'the home in which she had sheltered and fed another's child' (see Ellman et al 1991 1324, see Superior Court of Orange County, Nos. X-633190 and AD-57638). On that reasoning, he granted the genetic parents full parental rights. Indeed, Judge Parslow based his decision explicitly on such things as the high heritability of I.Q. and other behavioral and physical characteristics (see Superior Court of Orange County, and see also Krim [1996] for discussion). As this case worked its way to the California Supreme Court, several courts in between reaffirmed the emphasis on the genetic. For example, the Court of Appeals noted that:

> As evidence at trial showed, the whole process of human development is 'set in motion by the genes.' There is not a single organic system of the human body not influenced by an individual's underlying genetic makeup. Genes determine the way physiological components of the human body, such as the heart, liver, or blood vessels operate. Also, according to the expert testimony received at trial, it is now thought that genes influence tastes, preferences, personality styles, manners of speech and mannerisms.
>
> (Cal.Rept. 286, 1991, 381)

The California Supreme Court justices did not rely explicitly on such claims; in fact, they noted that each woman in the case had submitted acceptable proof of maternity under California law – in Crispina Calvert's case, blood and genetic testing, and in Anna Johnson's case, proof of having given birth to the child in – and that there is 'no clear legislative preference' in civil code for one form of proof over the other (*Johnson* v. *Calvert* 5 Cal. 4th 92). But even here they note that it is reasonable to argue that 'while gestation may demonstrate maternal status . . . it is possible that the common law viewed genetic consanguinity as the basis for maternal rights' and that 'under this interpretation gestation simply would be irrefutable evidence of the more fundamental genetic relationship' (*Johnson* v. *Calvert*, quoting Hill, 92–93).

This same line of reasoning was applied in *Belsito* v. *Clark*, where the judge also noted that proof of giving birth used to be an acceptable proof of being a 'natural' parent because 'for millennia, giving birth was synonymous with providing the genetic makeup of the child that was born' (*Belsito* v. *Clark* 67 Ohio Misc. 2d 59), that is, giving birth was proof of the existence of the more fundamental genetic relationship. In this case, the court ruled in no uncertain terms that 'the law requires that, because Shelly Belsito and Anthony Belsito provided the child with its genetics, they must be designated as the legal and natural parents' (58). The ruling continued:

> there is abundant precedent for using the genetics test for identifying a natural parent . . . The genetic parent can guide the child from experience through the strengths and weaknesses of a common ancestry of genetic traits . . . [the genetic test] should remain the primary test for determining the natural parent, or parents, in nongenetic-providing surrogacy cases.
>
> (64)

While this kind of reasoning is far from being universally accepted in legal contexts, the fact that such arguments can get made *at all* suggests that at the very least a strong emphasis on the genetic relationship as defining parenthood. While this reasoning is made obvious in cases of contract pregnancies (surrogacy arrangements), it is of course applicable in cases of 'ordinary' pregnancies as well. And of course, the extension of this kind of reasoning to ordinary cases is what makes so tempting the idea that having a 'child of one's own' *demands* that there be a genetic relationship. If these kinds of claims are right, we cannot expect to be able to 'guide' adopted children through their development as successfully as we can guide our 'genetic' children, because we won't share the same genetic 'strengths and weaknesses.' Nor will an adopted child, unlike again a 'genetic' child, be likely to share our temperament, our intellectual abilities and interests, our personalities, or any of the other features that make us the people we are.

3. THE ILLUSION OF CONTROL

But to what extent do we really expect, or should we expect, our children to be like us? Physically, it is quite clear that children tend to resemble their (genetic) parents more than they resemble randomly chosen humans, and its reasonably clear that they tend to physically resemble their genetic parents more than they do random individuals chosen from among their particular social/geographic locality. But quite often, even when children physically resemble their parents more than they do the population in general, they still don't resemble their parents *very much.* Having a child that shares (half of) your DNA does not guarantee that the child will look very much like you; obviously, children often differ physically from their parents in ways we find significant. This is not to say that being the genetic parent of a child doesn't alter the probabilities of particular kinds of phenotypic outcomes – clearly it does. If it didn't, some of the cases of the 'wrong' embryo being implanted in IVF procedures would, very likely, not have come to light. But the IVF cases are striking in large part because the only cases where the (putative) parents became aware that something was amiss on the basis of their children's physical appearances have been those in which the children born have been 'the wrong' color (see Noble-Allgire, 1999; Liebler, 2002). Indeed, the few studies there have been of error rates in IVF and similar assisted reproductive technologies would seem to imply that there are currently *many* parents happily raising children they have no genetic relationship to and are simply unaware of the fact; their children do not fail to resemble them sufficiently for them to even be suspicious (see Noble-Allgire, 1999; Fischer, 1999). This should come as no surprise; recall the cases of parents whose babies were 'switched at birth' and who only discover much later – and for reasons that have little or nothing to do with physical appearance – that they are not the birth parents of the child in question.

With respect to (broadly) psychological traits – temperament, intelligence, creativity, etc. – those who support the importance of genetic parenthood are on even shakier ground. Again, it is clear that the genetic children of parents resemble their parents on many of these psychological traits more than they do people from the population at large, and despite the difficulties disentangling genetic from environmental influences it would be foolish to suggest that none of the correlation between parents and children is associated with their shared genetic (or more broadly biological) heritage (see e.g. Plomin, DeFries, and McClearn 1990 for review). But statistical correlations aside, siblings generally do not share identical – or even terribly similar – temperaments with each other nor with their (genetic) parents (see Lewontin, Rose, and Kamin, 1984). Even identical twins are not the same people; they can, and often do, differ in a variety of psychological traits.

Indeed, some authors have noted that the most striking feature of sexual reproduction in humans is the *unpredictability* of our children; Wheeler notes that 'even those who embark upon parenting in the most considered and controlled and rational way possible *do not know who is coming*' (Wheeler, 2001, 117, emphasis in original). And, Wheeler continues, for all our attempts to shape our children, through carefully chosen and created environments, children 'from birth both absorb and resist our influence' (Wheeler, 2001, 118). Indeed, our inability to reliably shape our children's personalities, interests, etc., is often cited as evidence for these features having a biological basis (see Cohen, 1999); while this argument is conceptually confused, it does point towards what is perhaps the most important feature of children's development. *Neither* the child's genetic (and other biological) endowment *nor* the environment of the child are sufficient for us to understand – let alone predict! – even the broadest of a child's psychological features.

This should hardly come as a surprise. Development is fiendishly complex, so complex that the very idea of there being 'genes' that can defined or identified independently of the ways that they are used in development can be legitimately questioned (see Stotz forthcoming). What constitutes a 'gene', and what that 'gene' does, depend critically on the cellular environment surrounding the genetic material, and that depends critically on the developmental environment more generally. This matters to the rhetoric surrounding genes – while it is now often claimed that the Human Genome Project, and related enterprises, have revealed that humans have only between 25,000 and 40,000 genes, with the current rough consensus being somewhat under 30,000 (see Southan, 2004), this claim requires an interpretation of 'gene' that seems conceptually confused. When one takes account of the different ways that particular stretches of DNA can be 'transcribed' and used in protein production and the regulation of other developmental pathways, it becomes clear that each of the 'genes' identified in the standard counts of gene-number can in fact be involved in the production of *many* different proteins and the regulation of *many* different molecular pathways (see e.g. Roberts and Smith, 2002).

In addition, these processes of gene-regulation and protein production depend critically on the local intracellular environments encountered (see e.g. Xu, Modrek, and Lee, 2002). Genes, however one picks them out as molecular entities, are therefore not best thought of as a set of 'blueprints' nor as a 'recipe' for development; rather, they are best thought of as simply one kind of resource used in development, where the use to which they are put depends critically on the other resources present in development and they way *those* resources are currently being used (see Moss, 2003 for discussion).

The effort to count the number of human genes does provide us with one important lesson, however. The estimated number of human genes in the human genome is roughly the same as the number of genes in the mouse genome (see Boguski, 2002), and only about twice that of the fruit fly (see Baltimore, 2001). Nor is there anything startlingly unique about the particular kinds of genes that humans have compared to those that mice have; we not only share about the number of genes as mice, but most of the genes themselves are very similar as well (see Boguski, 2002). What follows from these low estimates of the number of human genes and the similarity of the genes in humans to those in other vertebrates is, I want to suggest, that it is *not* our unique 'genetic endowment' that makes us human. There are, by the standard counts, simply too few genes, and those genes that there are turn out to be much too similar to the genes of other mammals, for this to make sense.

While obvious, the rhetoric of the genome project makes this point hard to see. For example, in one article, Baltimore claims *both* that sequencing the human genome has provided us with 'the information needed to create a human being' (2001, 814) *and* that the low number and highly conserved nature of the 'genes' discovered in humans points towards 'what gives us our complexity' being 'a problem for the future' and *not* one that will be solved by attention to our genes' (2001, 816)! The 'information needed to create a human being' is *not* encoded in our genes. Indeed, insofar as we choose to think about development in terms of information, the information needed for development is actively *created* during development by the interaction of various kinds of resources used during development, including genes, proteins, intracellular membranes, and the like (see Oyama, 1985/2000; Moss, 2003). It makes no sense to think of DNA as encoding 'information' in any way that is different from the way that most other aspects of the developmental process 'encode' information (see Oyama, 1985/2000; Oyama, Griffiths, and Gray, 2001; Moss, 2003).

The complexities of development, including the highly complex and variable ways in which particular stretches of DNA get used in development, have taught us that is very difficult to link particular stretches of DNA – or any other particular conception of a 'gene' – to particular developmental processes that will result in a particular physical or behavioral trait. So it would be very foolish of us to expect that our genetic relationship to our children will result in their being very much like us. Nor ought we to expect that the fact that we share a certain percentage of our 'genetic heritage' with our children will give us very much of an advantage in understanding our children's unique strengths and weaknesses. Sharing genes does not, in fact, count for all that much.

So, in short, the confidence that parents have that a 'child of their own' will resemble them in key respects, or will have personalities or physical characteristics that are 'correct', is deeply misguided. The importance currently placed in having a 'child of one's own' emerges not from plausible interpretations of biological research showing the importance of genes to development, but rather from a misplaced confidence in the importance of genetics more generally. And that confidence, as misplaced as it is, emerges quite naturally from the over-stated claims and inflated rhetoric of the researchers involved in (human) genetics. The attempt to control the characteristics of one's children by ensuring that they are one's own (genetic) children is, in other words, premised on a confusion about the relationship between a shared genetic heritage and the complexities of development.

Obviously, genetic (and other forms of prenatal) testing (via, say, amniocentesis, etc.) fit into this vision of an attempt at control as well. Even more than the desire to exclude 'damaged' children, we can view these tests as an attempt to find out, as Wheeler put it, 'who is coming'. But Wheeler is right in noting that we can't in fact know who is coming – knowing that one is pregnant with a male or female, knowing that the child lacks certain genetic markers known to be associated with various diseases – none of this tells us what our child-to-be will be like, nor even that they will be healthy; too many diseases and conditions are idiopathic and cannot be revealed by contemporary testing technologies. So while it is true that testing can (almost) exclude some conditions, such as Down's syndrom, as there are far more kinds of developmental abnormalities than can be currently tested for, testing cannot guarantee the absence of a disorder. To take another example, although ultrasounds (sonograms) are known to be of at best questionable usefulness in the case of 'normal' pregnancies, they have become part of a ritual that generates confidence that the pregnancy is progressing 'as it should' *despite* their ineffectiveness as a diagnostic tool (see Stephens, Montefalcon, and Lane, 2000). For similar reasons, it is reasonable to argue that family histories are somewhat over-rated – most of our medical problems are not related in any clear-cut way to the medical problems of our immediate ancestors. We are, for example, far more likely to die from such things as infectious diseases, accidents and injuries, idiopathic or environmentally induced cancers, etc., than we are to die from a straightforwardly heritable condition like Huntington's disease (see WHO 2003 Statistical Annex for details).

None of this is to deny that there are sometimes advantages to knowing something about one's biological ancestry – again, it is undeniable that people whose ancestors came from particular local populations have different probabilities of carrying genes that are generally regarded as dangerous or at least as potentially problematic in contemporary contexts. To name a few of the well-known examples, people descended from 'Caucasians' from western Europe are more likely to carry genes associated with cystic fibrosis than the population at large, people descended from populations with endemic malaria are more likely to carry the HbS 'sickle cell' genes than the population at large, and women of Ashkenazi ancestry are more likely to carry the BRCA1/2 genes associated with breast cancer. And of course, knowing that a close (genetic) relative suffered from e.g. cystic fibrosis or had sickle-cell anemia, etc., does give one some information about one's own (or one's genetic children's) chances of having related problems.

But these considerations are at best tangential to the arguments that people generally give for wanting 'children of their own'. No one, I take, thinks it terribly important that their children resemble them with respect to the likelihood of their contracting various diseases – indeed, the whole *point* of e.g., prenatal genetic testing is generally to insure that one's children *won't* have the same likelihood as their parents had of contracting various diseases! The worry about adoption, in this case, is not, I take it, that your adopted children won't resemble you with respect to disease risks generally, but that they will have a genetic condition you won't find out about until 'it is too late' (too late to not adopt them, one supposes); this is, again, primarily a concern about getting a child that is 'damaged' in some way (again, see The New York State Task Force, 1998, 87). But again, even prenatal genetic testing combined with e.g. ultrasound examinations and the 'best' prenatal care recommended does not guarantee a healthy baby, nor a child that will never have medical problems related to their 'genetic' (or other biological) endowments. Indeed, except in very specific circumstances – where, for example, a particular risk has already been identified on the basis of family history – such testing does not markedly change the *probability* that one will have a healthy baby (see e.g. Bucher and Schmidt, 1993). There is even some evidence suggesting that prenatal care is less efficacious than has generally been thought in increasing the probability of one having a healthy child – that, indeed, 'standard' prenatal care is essentially useless by any reasonable empirical measures (see Alexander and Kotelchuck, 2001; Huntington and Connell, 1994).

In the next section, I suggest that once the lack of control inherent in raising *any* child is accepted, and the false lure of genetic similarity is rejected, there is ample room to reevaluate what it means to have a 'child of one's own' in ways that do not privilege having a particular genetic relationship to the child in question. Even if these arguments are accepted, the process of moving, as a culture, towards a view of parenthood that is not wrapped up in the illusion of biological control will no doubt be a long one. But the sooner we begin down that road, the better.

4. PARENTHOOD, ADOPTION, AND CHILDREN OF ONE'S OWN

Tamara Pyles and Marla Liston, who had been in a relationship for over fifteen years, decided to raise a child together, and decided that Pyles would be the biological mother. Three years after the birth of the child, Connor, their relationship ended. Liston brought legal action against Pyles, arguing that as one of Connor's parents she should be granted visitation rights and shared parenting responsibilities (No. 97APf01–137, 1997 Ohio A Lexis 3627). In 1997, however, the court ruled that because Liston was not Connor's 'natural' parent, and had not legally adopted Connor, that she had no legal rights nor responsibilities with respect to Connor (Lexis, 3627, 8). Insofar as one thinks that being a parent is primarily about having a particular biological relationship to the child in question, this obviously makes sense. But at the appellate level, one judge dissented from the opinion described above. Judge Tyack argued that the 'facts before us indicate that Connor was indeed the appellant's child as well as the appellee's' (Lexis, 3627, 32). His reasoning is that:

> The facts in the record before us indicate appellant was actively involved in the decision to bear this child. Appellant nurtured this child and provided for this child's well-being ... This was a sixteen-year, committed relationship between two people that resulted in the birth of a child ... The birth was a planned one, the purpose of which was to extend an already existing family unit. After the birth, the child was cared for by both parties. Wills and a trust fund were modified and/or formed by both parties to provide for the new addition to their family.
>
> (Lexis, 3627, 37)

For Judge Tyack, being a parent was not a matter of being in a certain *genetic* or biological relationship with a child, but a certain *social* relationship. And the majority opinion's concentration on the genetic, Tyack implied, had the effect of denying the importance of that social relationship.

Judge Tyack's stress on the social, it seems to me, gets the matter exactly right. If one accepts the view he expresses, then interpretations of parenthood that highlight the importance of a shared genetic heritage will seem wrong-headed. The idea that 'while gestation may demonstrate maternal status ... it is possible that the common law viewed genetic consanguinity as the basis for maternal rights' and that 'under this interpretation gestation simply would be irrefutable evidence of the more fundamental genetic relationship' (*Johnson* v. *Calvert*, quoting Hill, 92–93) can be seen as getting the matter backwards. Gestation is not an evidential proxy for the more fundamental 'genetic' relationship; if anything it is that the 'genetic' provides evidence of there having been a 'fundamental' relationship (sexual, in this 'usual' case) between the putative parents. This conclusion is not absurd even from a straightforward legal point of view – in fact family law texts note that some 'modern courts occasionally permit juries to find defendants liable [in paternity suits] despite serological evidence of no paternity' (Ellman et al. 1991, 894). This seems to suggest that even in the U.S. context a conception of parenthood that looks towards personal and social relationships rather than towards biological relatedness is not impossible.

To take a brief cross-cultural detour, the early cultural anthropologist Malinowski claimed that the Trobriand Islanders he studied saw no biological connection between children and (living) men (see Malinowski, 1916, 221). While we may be legitimately suspicious of claims like these, as early anthropological reports often underestimate the sophistication of the respondents's beliefs, Malinowski did present compelling evidence that among those islanders he studied, 'fatherhood' was about the relationship that the man has to the woman who gives birth to the children. Aside from several incidents where men returned after years of absence and accepted children just born as their own (with all the appropriate joy at having a child) (Malinowski, 1916, 223–224), his informants were quite clear that the *only* way of a child being 'illegitimate' was by being born to an unwed woman (Malinowski, 1916, 222). The point of this is not to hold up the Trobriand Island culture as an exemplar, but rather to make clear that there is nothing necessary about the views of parenthood that North American culture has adopted or which it may yet adopt in the future; different cultures have done things differently, and if there are good reasons to revise our current practices, it would at least be worth thinking about doing so.

So why might we revise our current practices? I have already suggested part of the answer – our belief that a child related to us will be like us in important or significant ways is off the mark. Thinking that sharing a genetic (or other biological) heritage with our children makes them 'ours' fails to take seriously enough the unpredictability of development. Neither by giving our children 'the right' genes nor by giving them 'the right' environment can we ensure that they will become the people we wish them to be. And while it is impossible to ensure that we get the children we want, we ought to consider seriously whether the *desire* to get the children we want is a desire we ought to affirm. That is, while we can't ensure that we will get the children we want, perhaps we ought not even *want* to be able to control the kinds of children we will get.

In 'A Case Against Cloning' Gilbert Meilaender critiques reproductive cloning on the grounds that reproductive cloning would 'give a new seal of approval' to our 'despotic' tendency to try 'to make children after our own image' (2001, 80–81). Meilaender argues that sexual reproduction by its very nature acknowledges 'the limits . . . of our control' (2001, 80), but, as noted above, this underestimates people's (misplaced) faith in genetics to determine outcomes. Demanding that the only children who are properly yours are those children who share part of your genetic heritage is itself demanding that you be able to maintain a kind of control over your offspring. In other words, Meilaender's concern about control can be extended to encompass those people who only want 'their own' (biological) children because they fear that 'other' children might be genetically inferior (or at least too dissimilar) or already 'damaged' by their developmental environments. In all these cases, we are trying, in Meilaender's words, to produce 'children to suit our aims and purposes' (2001, 81) rather than accepting them as their own, unique, people. But, as Wheeler reminds us, parenting is an inherently 'risky enterprise' in which our 'weighty sense of responsibility' is juxtaposed to the 'frustration of helplessness' (Wheeler, 2001, 118); we ought, both Wheeler and Meilaender suggest, to be willing to embrace that risk and our essential inability to control who our children will be and what they will become.

Admitting that there is substantial uncertainty and risk in raising a child, whether we are genetically related to the child or not, and accepting the fundamental lack of control we have over who, in the end, our child will become, can point us away from a desire to have children that are related to us genetically and, perhaps, towards a position that accepts adoption as an equally valid way to form families. This would require a substantial change, as there are currently some very strong cultural biases against adoptions (at least in the United States). Elizabeth Bartholet documents some of these in her book *Family Bonds* (see Bartholet, 1993). The most serious, perhaps, is that while the right to reproduce *biologically* – to have 'children of one's own' through 'ordinary' sexual reproduction or through any one of a number of Assisted Reproductive Technologies – is considered sacrosanct in contemporary western cultures (see The New York State Task Force, 1998, especially Chapter 6), having a child by other means is, far from being a right, often considered somewhat suspect. So, for example, Bartholet notes that, through parental screening in adoption cases, the adoption agencies decide who gets to be parent at all (who gets 'disqualified' from adopting), and further, by rating those who are deemed 'qualified' to be parents, which parents will get which available children and how long those potential parents will have to wait (see Batholet, 1993, 33). Batholet suggests that by subjecting people who wish to adopt to such scrutiny, but not subjecting people who wish to reproduce biologically to similar government interference, society sends the message that while 'it trusts what goes on when people give birth and raise a birth child' it also 'profoundly distrusts what goes on when a child is transferred from a birth to an adoptive parent' (1993, 34). These, and the many other ways in which adoption is made difficult, time-consuming, inconvenient, and often very expensive, signal that (again, at least in the U.S.

context) being a parent by adoption is a 'poor second best' to biological parenthood (Bartholet, 1993, 34).

Bartholet argues that this situation is unfortunate – she notes that 'it makes no sense for a society that thinks of itself as sane and humane to be driving people in the direction of child production rather than adoption' (1993, 35). And there is no denying that this is exactly what contemporary society does – aside from the long and costly parental screening process in adoptions, prospective adoptive parents are faced with a 'multi-tier' system. It is relatively inexpensive to adopt a 'special needs' child through a state agency, though adopting through such government agencies results in a parental screening process that is likely to be much more invasive than would be the case in adoption through a private agency (see The New York State Task Force, 1998, 87–91). Adopting 'healthy infants' through state agencies, though, is very difficult; private adoption agencies and related businesses, including of course agencies and attorneys specializing in international adoptions, provide better access to infants, albeit at much higher costs (see National Adoption Information Clearinghouse , 2004). Current adoption policies seem to merely replace the desire to control the process of parenthood through biological means with the desire for non-biological parenthood to be controlled and controllable, both by the government agencies involved and by the prospective adoptive parents.

The focus on biological relatedness as the *sine qua non* of the parent-child relationship damages adoption in other, more subtle ways as well. One of the fears often cited by potential adoptive parents is the fear of legal or personal problems with the birth mother (see New York State Task Force, 1998, 90–91); this fear is often exacerbated in the case of 'open' adoptions, where the birth mother is known to and knows the adoptive parents. But these fears cut both ways – the distrust of adoption and misplaced faith in the importance of biological relatedness make birth mothers less likely to consider adoption and government agencies less encouraging of adoption as an option in cases where children need to be removed from the homes of their biological parents (see Bartholet, 1999). It is possible that many more people would consider placing their biological children up for adoption if the social stigmas associated with adoption were reduced or eliminated. Indeed, it seems overwhelmingly likely that to a large extent, the so-called 'loss' claimed to be experienced by people who don't know their 'genetic offspring' is *entirely* the result of the social focus on biological relatedness determining what it means to be a parent. Reducing the emphasis on the biological might well reduce the desire some people claim to have to 'find' their biological parents or their biological offspring and hence the fears associated both with adopting a child and with putting a child up for adoption (see Fischer, 1999; Bartholet, 1993).

Recognizing that parenthood can – and should – be about a particular set of social relationships rather than about biological relationships, could go a long way towards ameliorating these problems, and would therefore be a boon both to many of the children currently in foster care or homes that are problematic in a variety of ways, as well as to potential parents wishing to adopt. With the stigma associated with adoption reduced, there would, it seems, be increased opportunities for the acceptance of alternative ways of being a parent. This change could, it seems likely, benefit all the parties in adoptions, including the biological parents.

Unfortunately, the kind of system described above is a long way from being implemented. The bitter irony inherent in current attitudes towards adoption and the current systems in place in the U.S. is that while current policies *impose* a kind of control that is *in fact* lacking from 'biological' parenthood, adoption is still considered to be more 'risky' and less well-controlled than is 'biological' parenthood. A healthier, or at the very least more realistic, attitude towards parenthood would acknowledge and embrace the lack of control and real risks we take in parenting. Children are not, in the end, 'ours' to mold or to shape. At best, we can help children become their own people. And we can do that with children that are genetically unrelated to us as easily as with children that are biologically 'our own'. Reshaping public policy to recognize and acknowledge this would be a good start.

ACKNOWLEDGEMENT

I would like to thank Sharyn Clough for her help developing the ideas in this paper, as well as for reading and making helpful suggestions on several drafts of the paper.

REFERENCES

Alexander GR, Milton K (2001) Aessing the role and effectiveness of prenatal care: history, challenges, and directions for future research. Public Health Rep 116(4):306–316.

Arneson RJ (1992) Commodification and commercial surrogacy. Philos Public Aff 21(2):132–164.

Baltimore D (2001) Our genome unveiled. Nature 409:814–816.

Bartholet E (1993) Family bonds: adoption and the politics of parenting. Houghton Mifflin, Boston, MA.

Bartholet E (1999) Nobody's children: abuse and neglect, foster drift, and the adoption alternative.Beacon Press, Boston, MA.

Bender L (2003) Genes, parents, and assisted reproductive technologies: arts, mistakes, sex, race, & law. Columbia J Gend Law 12:1–76.

Boguski MS (2002) The mouse that roared. Nature 420:515–516.

Breitowitz Y (2002) What's so bad about human cloning? Kennedy Inst Ethics J12(4):325–341.

Bucher H, Schmidt J (1993) Does routine ultrasound scanning improve outcome in pregnancy? Meta-analysis of various outcome measures. B M J 307:13–17.

Cohen DB (1999) Stranger in the nest: do parents really shape their child's personality, intelligence, or character? John Wiley & Sons, Inc, New York, NY.

Dawkins R (1987) The blind watchmaker. WH Norton, New York, NY.

Ellman IM, Kurtz PM, Bartlett KT (eds) (1991) Family law: cases, texts, problems, 2nd edn. The Michie Company, Charlottesville, VA.

Fischer JD (1999) Misappropriation of human eggs and embryos and the tort of conversion: a relational view. Loyola of Los Angel Law Rev 32:381–429.

Gilbert W (1992) A vision of the grail. In: Kevles D, Hood L (eds) The code of codes. Harvard University Press, Cambridge, MA.

Hood L (1993) Biology and medicine in the twenty-first century. In: Kevles D, Hood L (eds) The code of codes. Harvard University Press. Cambridge, MA.

Human cloning foundation (1998) The benefits of human cloning. Internet http://www.humancloning.org/benefits.htm

Huntington J, Frederick AC (1994) For every dollar spent – The cost-savings argument for prenatal care. N Engl J Med 331(19):1303–1307.
Hurwitz I (2000) Collaborative reproduction: finding the child in the maze of legal motherhood. Conn Law Rev 33(Fall):127–180.
Kirby MK, Patrick HH (1998) Evaluating older pre-adoptive foster children. Prof Psychol-Res Pr 29(5):428–436.
Krim TM (1996) "Beyond Baby M: International perspectives on gestational surrogacy and the demise of the unitary biological mother." Annals of Health Law, 5:193–226.
Lewontin RC, Rose S, Leon JK (1984) Not in our genes: biology, ideology, and human nature. Pantheon Books, New York, NY.
Liebler R (2002) Are you my parent? Are you my child? The role of genetics and race in defining relationships after reproductive technological mistakes. DePaul J Health Care Law 5(Summer):15–56.
Malinowski B (1916) Baloma: Spirits of the Dead in the Trobriand Islands. Journal of the Royal Anthropological Institute 46:354–430.
Mayr E (1982) The growth of biological thought: diversity, evolution, and inheritance. Harvard University Press, Cambridge, MA.
Meilaender G (2001) A case against cloning. In: Cole-Turner R (ed) Beyond cloning: religion and the remaking of humanity. Trinity Press International, Harrisburg, PA.
Moss L (2003) What genes *can't* do. The MIT Press, Cambridge, MA.
National Adoption Information Clearinghouse (2004) A service of the children's bureau, administration for children and families, U.S. department of health and human services. Online at: http://naic.acf.hhs.gov For costs, see http://naic.acf.hhs.gov/pubs/s_cost/index.cfm.
Noble-Allgire AM (1999) Switched at the fertility clinic: determining maternal rights when a child is born from stolen or misdelivered genetic material. Miss Law Rev. 64 (Summer):517–594.
Oyama S (1985) The ontogeny of information: developmental systems and evolution, 2nd edn (2000). Cambridge University Press, Cambridge, MA.
Oyama S, Griffiths PE, Gray RD (2001) Cycles of contingency: developmental systems and evolution. MIT Press, Cambridge, MA.
Plomin R, DeFries JC, McClearn GE (1990) Behavioral genetics:a primer, 2nd edn. W.H. Freeman and Company, New York, NY.
Report of the high-level panel on threats, challenges and change (2004) A more secure world: Our shared responsibility. United Nations' Publication.
Roberts GC, Smith CWI (2002) Alternative splicing: combinatorial output from the genome. Curr Opin Chem Biol 6(3):375–383.
Roseman E (2004) Common fears in adoption. Adoption.com. Internet http://library.adoption.com/Adoption/Common-Fears-in-Adoption/article/1714/1.html
Sorosky AD (1995) Lessons from the adoption experience: anticipating times of developmental conflict for the ART [assisted reproductive technologies] child. Address before the annual meeting of the American Society for reproductive medicine (Oct. 8, 1995) (transcript on file with the Connecticut Law Review).
Southan C (2004) As the yo-yo stopped? An assessment of human protein-coding gene number. Proteomics 4(6):1712–1726.
Stephens MB, Montefalcon R, Lane DA (2000) The maternal perspective on prenatal ultrasound. J Fam Practice 49(7):601–604.
Stotz K In Press. 'With genes like this, who needs an environment? Postgenomics's argument for the "Ontogeny of Information".' Philosophy of science 73(5), proceedings PSA 04, December 2006. Preprint: http://philsci-archive.pitt.edu/archive/00002456/
The New york State Task Force on Life and the Law (1998) (April). Assisted Reproductive Technologies: Analysis and Recommendations for Public Policy New York. Stotz, Karola C. Forthcoming. With Genes Like That, Who Needs an Environment? Forthcoming in Philos Sci (Supplement: PSA 2004, Symposium Papers).
Wheeler S (2001) Contingency, tragedy, and the virtues of parenting. In: Roland Cole-Turner (ed) Beyond cloning: religion and the remaking of humanity. Trinity Press International, Harrisburg, PA.

WHO (World Health Organization) (2003) The World Health Report 2003:Shaping the Future. Available online at: http://www.who.int/whr/2003/en/

Xu Q, Modrek B, Lee C (2002) Genome-wide detection of tissue-specific alternative splicing in the human transcriptome. Nucleic Acids Res 30(17):3754–3766.

Legal Cases Cited

Belsito et al. v. Clark et al 67 Ohio Misc. 2d 54; 644 N.E. 2d 760 1994.

Johnson v. Calvert. 5 Cal. 4th 84; 851 P. 2d 776 1993; 19 Cal. Rptr. 2d 494. See also Cal.A 4th 977 1991; 286 Cal. Rptr. 369. See also Superior Court of Orange County, Nox X-633190 and AD-57638, Judge Richard N. Parslow, Jr.

Liston v. Pyles. No. 97APF01–137 1997, Court of Appeals of Ohio, Tenth Appellate District, Franklin County.

CHAPTER 12

IS A TRANSCULTURAL LAW FOR HUMAN GENETICS AND BIOTECHNOLOGY POSSIBLE?

CARLOS M. ROMEO-CASABONA

Professor in Criminal Law, Director, Inter-University Chair BBVA Foundation – Provincial Government of Biscay in Law and the Human Genome. University of Deusto and University of The Basque Country, Bilbao, Spain, cromeo@genomelaw.deusto.es

Abstract: In this contribution the author suggests the necessity to find a new focus to the law related to genetics and biotechnology. Phenomena like globalization are not equal to the process of universalitation of values and goods to be protected by law. For a juridical approach to these matters it seems necessary to explore first the scientific and cultural contexts involved. Transculturality is an acceptable way to try to find common values that could be relevant to elaborate a universal legal framework for genetics and biotechnology. Then there is framework for harmonization, but giving anyway priority to universal recognized human rights. Human rights also present an objective dimension that permits through them the protection of realities or situations independent of the possibility of accepting the existence of a subject titleholder of a specific right. The challenge nowadays is to be able to recognize to human rights also a collective perspective. For Criminal Law specifically there is needed to avoid the recourse to a merely symbolic Law, in reference to which the legislator is more worried about expressing a moral and social rejection towards certain activities and to calm society than for an effective legal persecution of the same (as some are still in the future)

1. A NEW SETTING FOR LAW IN THE FIELD OF HUMAN GENETICS AND BIOTECHNOLOGY: GLOBALIZATION

As a general rule, now are well known the real and virtual benefits, which can arise through research and other actions relating to the human genome and the innovations that such knowledge can generate in the field of biotechnology. As far as human biotechnology is concerned, its achievements are being focused in two areas of great importance for the human being: health (new diagnostic procedures and treatments) and reproduction, whether this is related or not to health problems of the couple or the future child.

A. Fagot-Largeault et al. (eds.), *The Influence of Genetics on Contemporary Thinking*, 181–193.

Nevertheless, the creation of procedures, the creation of biotechnological products and the investigation in which they are based must be compatible with the adoption of precautionary and security measures in the handling of live matter. Especially so when the live matter has been the object of genetic modifications, whose interference in other living beings, including the human being, are still unpredictable.

In the entirety of interests which can become criss-crossed in the complex tangle of human genetics and biotechnology, it has especial relevance for this contribution those which have a more immediate relation with some fundamental rights. We are interested, in particular with rights such as the right to life and the right to physical and moral integrity. These are being proclaimed by some international legal instruments and by the Constitutions of some States. In the legal systems, the importance of life, human integrity and, with some limitations, life and integrity of the unborn (embryo in vitro and the embryo or foetus in gestation) are recognized as essential juridical goods. On the other hand, human dignity, an individualistic axis in which pivot and are measured the human rights and the fundamental rights, can also be affected with the development of human genetics and biotechnology (thinking as an example, the use of human beings as guinea pigs while undertaking some clinical trials). However, the issue, which raises the greatest interest, and at the same time complexity, consists in detecting if new individual or collective juridical goods can be deduced, and what could be the sources to identify, deduce and derive them.

Therefore, these juridical goods must be protected with maximum intensity, a task that is assumed through Criminal Law, specifically against the most serious attacks that those goods can be subjected to. In any case, a relevant question is also whether they are sufficiently relevant and at the same time, concrete so as to merit have to recourse to legal-criminal instruments, always as *ultima ratio*, for its protection. From the viewpoint of a hypothetical intervention of Criminal Law as a control instrument for astraying practices related to human genetics and biotechnology, some characteristics must be set out. These can be decisive not only to evaluate the ‘whether to’ of the normative intervention but also the ‘how’ of such intervention.

Anyway, for a juridical approach to these matters it seems necessary to explore first in which scientific and cultural context are developing these matters.

Firstly, one must note the expansion and velocity with which new discoveries and applications are being produced in this field. This reflects the dynamism of the sector, the competitiveness that exists among different groups of researchers, but also the social perplexity, which is produced as seen by how many of those novelties test the foundations of well established social perceptions and values. This perplexity also points out the incapacity of traditional legal reasoning to offer, at least in all cases, efficient and/or calming answers. It has been denounced accurately that we are globalising profit and trades, but we are not globalising justice (Sulston, 2004, 32).

Secondly, the investigations which provide novelties in human genetics and biotechnology require neither infrastructures nor exceptional material means, as they are neither very costly nor difficult to obtain. The decisive factor is just the qualification of the investigator in the specific field. Potentially, this means that these activities could be undertaken in any country, as it would be enough for a small group of investigators to establish themselves backed with a minimum infrastructure and resources, regardless of the potential capacity of investigation of the welcoming country. This is a proof of that globalization also occurs in the field of human genetics and biotechnology.

Cultural peculiarities, in particular moral, religious and legal traditions entail relevant differences from some States to others when they deal with the juridical framework of human life sciences. Nevertheless, when are attached assessments to human genetics, which do not always connect solidly to previously well defined axiological principles, it is not uncommon that quick perception changes can happen and that even the policy-makers may supply mimetically imported answers. Paradoxically, these cultural divergences have been coexisting in some States with the assumption of an ideological pluralism by their citizens. In some cases, this pluralism has met strong resistance by some ideological or religious groups in order to re-examine concrete traditional assessments, especially those related with the respect and protection which human life in its different forms deserves. Resistance has also been met when confronted with new phenomena or realities, for example, in relation with critical moments of the beginning of human life (thereby the ethical and legal statute of the in vitro human embryo) and the end of human life (e.g. the decision to terminate a vital medical treatment).

This pile of axiological divergences or of ways of dealing with new situations made it more difficult to find meeting points over the acceptance or not of some human genetic novelties.

In fact, International Law has encouraged a global perspective in relation to human genetics and biotechnology. This global perspective has been benefited by the concurrence of several factors:

a) In State laws there were lacking ethical and cultural references of a clear and undisputed application to the new challenges which rose from human genetics and biotechnology. This means that there exist certain values related to human genetics and biotechnology (e.g. human rights, juridical goods) which have received universal recognition with greater ease than other rights, shall we say more traditional (e.g. civil and political rights as related to social groups). These latter ones were previously elaborated and have sometimes not had a correspondence with certain cultural and ethical beliefs of some human communities. In this respect, it is noteworthy that many countries which neither have a cultural tradition in these matters nor are at the forefront of biomedical investigation, have fervently adopted legal measures aimed towards, for example, establishing the informed consent as a basic right of the patients, or have been quick to prohibit human reproductive cloning.[1]

[1] E.g. Peru (Penal Code, art. 324), Vietnam and China, even though this last also has been quick to authorize the so-called 'therapeutic' cloning.

b) The initial surge and development of the Law of Human Genetics and Biotechnology has manifested itself generally as 'soft law', it is to say, as non-coercive law, and accordingly with that, without strong legal consequences. Probably, the most noteworthy exception to this tendency is in regard to human biotechnology, in which have been introduced in Comparative Law several offences whose punishment are, in general, very stiff (e.g., human reproductive cloning). This has led to a discussion of the purely symbolic effect that Criminal Law could irradiate to pursuit as a crime 'activities for the future', I mean, to apply penalties or other legal consequences to activities that are not still technically practicable. However, the Law on human and biotechnology has been moving in a slow and constant manner towards a Law characterized by rules of ascertainment, by its legal binding character, as they are backed with more frequency by sanctions or other legal consequences.
c) Subject matters proper of human genetics and biotechnology, as well as in general human life sciences, which have been the object of International Law, do not compromise State Sovereignty, but it is likely true also that we are beginning to contemplate something beyond the existing conception of the nation-state as a manifestation of globalization rather than internationalisation (see in this sense, Singer, 2004, 8).

The above-mentioned factors have maybe helped the international development of some bioethical principles, but they just have mainly helped a purely process of globalization of Bioethics and it shows that a lack of trans-cultural integration exists. In fact, this way of globalization is maybe not the best to ensure that it is truly a result of an achieved trans-cultural process. The challenge of our time is, therefore, to ensure that ethical and legal globalization materialzes in a trans-cultural framework that accommodates universal acceptance of certain shared values and rights capable of affording the answers required by the challenges of a globalised world.

None the less, before continuing with these thoughts, one must note that for Criminal Law the phenomena of globalization carry other specific aspects as they are: the globalization of criminal behavior, that is, the commission of criminal acts which surpass boundaries, the appearance of new types of criminal behavior and the difficulty for Criminal Law to be operative in terms of territoriality. In fact, since Criminal Law is still based on the principle of territoriality, that is, the application of law exclusively to violations in the own State territory,[2] this entails great limitations to be able to apply its legal norms further than its political–legal frontiers: judges have only jurisdiction to apply the laws of their State and within the limits in which it exercises its sovereignty. Therefore they can not persecute, at first, those persons who have committed punishable acts outside the territory and to those who are outside of it, without prejudice to the exceptions which are

[2] The principle of personality (the application of criminal law to the citizens even thought the crime might have occurred outside the national boundaries) continues to present its own problems, as occurs in Germany in relation to the inventions with human embryos. See Eser / Koch (2003).

allowable in both situations. Precisely, the exercise of sovereignty is little prone to waive its rights in criminal matters, as this field of law is that in which sovereignty is most significantly reflected. This is without prejudice to some attempts that try to overcome these limitations (e.g. the International Criminal Court). However, it does not seem at the present that the most serious human genetics and biotechnology related abuses are going to be heard by the court, maybe with the exception, of reproductive human cloning.

2. THE BASIS FOR A TRANSCULTURAL LAW IN THE FIELD OF HUMAN GENETICS AND BIOTECHNOLOGY[3]

There have been attempts to establish a transcultural ethical system trying to avoid at the same time the imposition of a dominant ethics over the rest (e.g. western ethics based on Christianity). This process of reasoning can be valid for a parallel analysis about a possible Law of human genetics and biotechnology with a transcultural basis.[4]

This objective is acquiring a growing interest, given the accelerated process of economic and technological globalization and due to the risks which humankind is being subjected to, including the ethical and legal dimensions, and in particular Bioethics. Therefore, to find transcultural fundamental principles of Ethics and Law is a task which can not be delayed, especially if we take into account that other recent attempts in the human biotechnology sector have not had the desired success.

It has been noted that the task of creating a transcultural Bioethics and a Law of human genetics and biotechnology with a universal acceptance carry the risk of falling into a form of cultural imperialism. This could be derived from the established fact that throughout history, every society has had a tendency to impose its cultural fundaments and its ethical beliefs upon the rest (Singer, 2004, 106). It has also been noted that one must avoid the risk of falling into an individualism or collectivism. These are expressions used to make reference to the fact that in reality the individuals are conditioned in their cultural education by the community and the social surrounding in which they are, beyond abstract principles. On the other hand, the waiver to a certain universalisation also implies the non-desired risk of remaining in a relativism in respect to the rights related to human genetics and biotechnology, thereby one could not identify nor construct values and principles of universal acceptance.

Among the several hypothesis which have been used to base this transcultural Ethics and Law (Vallespin, 2004, 111) about which I can not stop to discuss in this presentation, I would like to highlight that which upholds that in order for a

---

[3] I have attempted to offer previously this focus (2004).

[4] I don't intend in this contribution to reach a definition of transculturality as referred to ethics or law, but for the purposes of it may be enough clear to mean that it focuses the minimal common traditions and values shared worldwide.

transcultural ethics to be acceptable, it should take as a starting point the elements or common ethical denominators which are present in the diverse cultures in the planet.

Some authors have wanted to establish this minimal common point in the principle of reciprocity. In fact, for Singer the principle of reciprocity should be the *Golden Rule*: "treat others as you would like them to treat you". According to Singer (2004, 141–143), the principle of reciprocity can be found with diverse formulations in all cultures, present and historical, and in all religions. However, neither in history (e.g. the doctrine of Jesus Christ, who preached to do good to those who do evil, when he exhorts to place the other cheek so that it can also be struck), nor in some current fundamentalist trends (e.g. radical movements based on religious inspiration, as Singer himself acknowledges) can one notice its acceptance or its respect, even though it responds, probably, to opposing motives. Furthermore, this principle of reciprocity as unifying force of a universal ethics seems at the present time to be very poor, though it still certainly maintains its interest at the present, in sight of the cultural riches reached by our civilization, in which one can notice several shared values.

It must be emphasized the uniformity role which has been undertaking, since decades ago, the construction of human rights, which being a creation of the western culture, has been accepted, some times for better and some times for worse, by people of other cultures of non western States. Specifically, in relation to human genetics and biotechnology, there exist in this sense, recent contributions which should stimulate that unifying process, such as the UNESCO Universal Declaration of the human genome and human rights (1997) and the Council of Europe's Convention on Human Rights and Biomedicine (1997) and the corresponding protocols of the last. In a sense, these legal instruments have contributed towards the globalization of Bioethics and it will be increased when the elaboration of the UNESCO Universal Declaration on Bioethics and Human Rights (2005) will be accomplished.

The theory of human rights has an ethical background of great importance and has an undeniable value capable of universalisation, as seen by the international acceptance that the human rights have obtained in International Law, some of which have reached a universality not open to discussion.

However, in spite of its importance, we must be aware that they are not always shared by all cultures, and that even in western culture, some of them are barely making their way, at least as legal regulations. In fact, the construction of human rights has been a matter of criticism, as it brings to an exaltation of the individual, which is a main characteristic of western culture. This is set against the collective world view of other cultures (this, as seen generally in the Far East, in some parts of Africa and in the indigenous populations of Central and South America), which with a holistic focus or other manner maintain that the creation of harmony in the community is possible starting from the obligations that the individual contracts with its members and vice versa. It is from the perspective of the duties of the community and with it that one will attain the respect of its members. Nevertheless, the unstoppable extension and penetration of the phenomenon of globalization requires us to find counterweights and balances against the very grave risks that could be derived from all embracing sources of power which can neither be controlled by neither the States nor the International Community with their actual resources. Precisely, human genetics and biotechnology is one of the most attractive temptations to try to surpass any limit, any control, and human rights can be an instrument suited for it, at least as a first step.

Be as it may, one must take advantage of the universal acceptance which human rights have been enjoying and must continue to prudently take them as a reference point to universally identify, assume and share a set of legally based ethical values. Furthermore, human rights are not static, nor they aspire to create a closed universe. On the contrary, they are always in constant evolution, taking in new rights in relation to human needs, and therefore constitute a very valuable instrument for the shaping of new rights in the context of genetics and biotechnology.

It follows that although in this presentation they can only be merely enumerated, the challenge of our time is to assure that an ethical globalization will happen in a framework of transculturality. This framework should welcome the universal acceptance of certain shared values and rights which could provide the answers which demand the challenges presented by a globalised world. To continue along this path, it is necessary to be able to conjugate both dimensions of principles and rights, that is, the individual with the collective, which should be constituted or reinforced as an axiological instrument and of good fellowship for the next decades.

Although the list would be never ending, I am going to limit myself to mention those that I consider essential for the scope of a transculturally based globalization. Therefore, I will dispense with the reference to certain civil and political rights, including some social rights, recognized by Universal Declarations or Treaties. I will also dispense with those which more specifically constitute the core of rights related to human genetics, as are: the human genome as being the heritage of humanity,[5] the right to self integrity, the right to ones genetic identity, all as a characteristic of the specie, as well as the right to the protection of personal genetic data, the right not to be discriminated on grounds of the genetic characteristics (UNESCO, Declaration, 2003, art. 11).

Furthermore, it is my understanding that at least the following principles could set a minimum basis for a universal recognition: the principles of responsibility (Jonas, 1984, 153), of solidarity (Sulston, 2004), of justice – whatever it could mean – (Rawls, 1971, 453), equity, tolerance (Kaufmann, 1999, 321 and also in Saada-Gendron, 1999), non-discrimination and responsibility towards future generations (Romeo-Casabona, 2002, 32). Notwithstanding, these principles and rights need for a collective perspective to be implemented adequately.

These principles and rights have the added value that they present individual as well as collective dimensions, or at least these can be deduced and implemented.

[5] In a symbolic dimension, as it was stated by the UNESCO's *Universal Declaration on Human Genome and Human Rights* (art. 1).

This double dimension has a relevant effect: they are rights that are preach able in individuals as well as in human groups and communities.

This proposal of looking at human rights from this double dimension, individualist and collective, wants to highlight that also in this manner the human groups and communities can be titleholders of rights. Additionally, the human being can be seen as titleholder of rights not only as an independent individual that is isolated from his surroundings, but also as part of the community to which he belongs. However, it must not be understood from this proposal that we are forfeiting the individualistic perspective of human rights, or that in the hypothesis of a conflict between the individualistic and the collective view of human rights, that the latter must prevail. The universal recognition that each individual is the titleholder of his own rights, unalienable and that cannot be waived must not be lost in any way. Specifically in the area of biomedical sciences the primacy of the human being over the sole interest of science or society should be assumed (see in this sense the art. 2nd. of the Convention on Human Rights and Biomedicine: 'The interest and welfare of the human being shall prevail over the sole interest of society or science'). According to this statement, priority is given to the former, which must in principle take precedence over the latter in the event of a conflict that could arise between them.

There is also some further clarification needed. Through the transcultural approach, as it is being presented in his contribution, we are looking for minimal common shared principles and for a universalisation of them as human rights and fundamental rights. But a complete unification of Law in the field of human genetics and biotechnology is not reachable. We should we aware that it should be kept a space for a pluralistic scope, which is respectful with the existent diversity of cultures and traditions. Then there is framework for harmonization, but giving anyway priority to universal recognized human rights (Delmas-Marty, 1997, 137).

I have left aside for a brief reflection the dignity of the human being. Among the various perspectives in which it could be shaped, there is no doubt that from it western perspective it is a sign of identity of the Kantian thought. Without prejudice to what other different perspectives from it have to offer, including the collective (as, in relation with the dignity of the people) it constitutes a paradigm of the protagonism of the superior worth of the individual in collective life. It is undoubted that is has been increasing in acceptance, now almost as a universal principle, without it being necessarily considered, as a general rule, as a fundamental right.[6] It is considered a quality inherent to the human being, which is projected legally over specific fundamental rights. Its relevance as a limit and sea wall against potential abuses from biomedicine in the human being is of first rate importance (Cortina, 2004, 29), as it carries with it the prohibition of utilising the human being – any human being – as an instrument and not as an end in itself. Unfortunately, the

[6] Notwithstanding, the German Constitution of 1949 states human dignity as a truly fundamental right (see art 1).

recourse to the dignity of the human being has been used excessively frequent and abusively in relation to the numerous advances in the biomedical sciences, using it against them as an authoritative argument, but not specifically founded to the issue being discussed. With this manner of argumentation, the possibility for a dialogue and for facilitating meeting and consensus points has been eluded. In spite of this easement, I consider that we must look deeper into the concept and content of the dignity of the human being. I think that still its contributions in the field of human biotechnology can be enriching, always when used with deliberation and loyalty to its true sense.

On the other hand, we should advance in other dimensions of the human dignity. In this sense, beside the assumptions in which one could have recourse to genetic interventions, some of which could consist, more than in an improvement, in interventions designed to select or bring about determined traits or biological characteristics considered desirable from a subjective point of view (e.g. of the parents in respect to their actual or unborn children, of those concerned, of the public powers). None the less, we should reflect on whether the personalistic conception of dignity is sufficient to contain the diverse assumptions which are associated to genetic interventions. These genetic interventions, at times, imply some conflicts which go further that the individual dimension of the human being, as collective dimensions can sometimes be affected, even the human species itself can be affected, at least from the perspective of a theoretical analysis which permits one to present this as a hypothesis.

It should really be thought over whether human dignity could not have, at the same time, a supra-individual dimension for such situations, starting from an objective dimension, of the assessments that human dignity can project. Some future measures of genetic intervention for achieving perfection or improvement (which if practised in the gametes before reproduction or in the zygote could become eugenic practices) could at least have the potentiality of affecting the human species or human ethnic groups. This would happen because the genetic make up which characterizes it as a specific species or as such a group would also be modified, and in this manner would also involve future generations. Independently of the specific individuals who could be affected, a different manner of conceiving human dignity, but that does not exclude others – in a supra-individual manner – could be used to slow down or reject such behaviors. Possibly, it could also provide more clarifying focuses when one has to determine what is 'normal' and what is 'pathological' (this question is presented by Canguilhem, 1966). On the other hand, the process of world globalization to which we are being witnesses of, probably also requires this type of elaboration in relation to human dignity. To this respect, one must bear in mind the Declaration of UNESCO on the responsibilities of the present generations towards future generations which establishes that "in full respect of the dignity of the human person and human rights, the human genome must be protected. Scientific and technological progress should not in any way impair or compromise the preservation of the human and other species" (Declaration of UNESCO, November 12, 1997, art. 6°).

3. DIFFICULTIES DEFEATS FOR A TRANSCULTURAL LAW IN BIOMEDICINE

An obvious example that this is not an easy task is shown, at the moment, by the works developed inside the United Nations with the object to approve a universal convention to prohibit both reproductive human cloning as well as the so called therapeutic cloning. In fact, reproductive cloning has noticeably met with an almost universal agreement about its prohibition. It has not been taken into account that there could be some scenarios in which its illegitimacy in the future is not so evident (e.g. if when it would be a really safe technique, one would use it to combat the infertility of a couple or with the purpose of preventing the transmission of hereditary diseases to the descendants; Romeo-Casabona, 1997, 21 ff. and Grupo de Expertos sobre Bioética y Clonación, 1999). While, at the same time, the miscalled 'therapeutic' cloning is facing irreconcilable cultural conceptions opposed over the respect and protection of human life in its onset and also to the pressure of the scientific community.

At the end the General Assembly of the UN has approved a Declaration. The ban addresses "all forms of human cloning inasmuch as they are incompatible with human dignity and the protection of human life". It also calls upon Member States to "adopt the measures necessary to prohibit the application of genetic engineering techniques that may be contrary to human dignity" (Unite Nations Declaration on Human Cloning, adopted on 8 March 2005). The vote of approval was 84 States in favor, 34 against and 37 abstentions and 37 absent States. In my view, this represents a clear example of a failure to reach a consensus. It is also an example of the juridical process of globalization, but lacking universal acceptance. Probably human cloning for research and therapeutic purposes is still far from having reached a general consensus (see further on Harris, 2004, 34 and Romeo-Casabona, 2002).

The dialectic tension, which the ethical and legal statute of the human embryo has generated, is also reflected in the difficult equilibrium, which is not free from contradictions, which tried to be maintained at the Convention on Human Rights and Biomedicine, where there was a ban of the creation of human embryos for use in experimentation, but, at the same time, admitting that they could be used for these ends, alluding implicitly to the excess embryos coming from the techniques of assisted reproduction.

4. THE INFLUENCE OF INTERNATIONAL LAW ON HUMAN GENETICS IN A RANGE OF NEW CONSTITUTIONAL RIGHTS OF THE "BIOETHICAL" CITIZEN

The everyday closer link between International Law and Internal Law has revealed itself especially in relation to human rights. It has been noted that in the immediate future it will be more intense each time in the specific field of the human life sciences. This observation is of an enormous importance; given that, for obvious reasons, up until recent times we could hardly find in the international legal instruments or in the internal constitutional laws explicit references to human rights affected by the recent scientific progresses, noting that this situation is undergoing a radical change.

Constitutional Law has a great potential, both as a receiver of human rights, which in a more specific manner are being involved by the human life sciences, as well as, as an instrument to resolve the conflicts that emerge from these. We can find numerous examples of conflict resolution in modern comparative constitutional law.

Undoubtedly, the incipient examples that exist over the acknowledgement of some rights associated with the human genome and the biotechnologies constitute a novelty for contemporaneous constitutional law. Moreover, this process, which has been slow in its beginning, is logical. If the human rights associated with these matters have settled inside International Law, it is logical that some fundamental rights could have a place in modern Constitutional Law, in a manner that they offer new perspectives for the protection of citizens. In this sense, as I had stated before there is no doubt about the influence which International Law is exerting over this budding Constitutional Law of Bioethics.

These rights have deserved acknowledgement including in the political constitutions of some States, as Switzerland (reproductive medicine and gene technology in the human field), Portugal[7] (personal dignity and the genetic identity of the human being) and Greece[8] (protection of person's genetic identity).

Regardless the respective content being on target, it must be recognized that these constitutional precepts constitute the first references to have this highest rank in relation to the concepts of autonomy of the individual, to the genetic heritage and to a right to a genetic identity of the human being as rights of the bioethical citizen (see Fagot-Largeault, 1985).[9] These comprise the core of a Constitutional Law of biomedicine that will be developing in the following years as a barrier against the pressures by certain researchers and enterprises that do not recognize any slow down against the progress of science and of economic benefits (the biocrats).

5. SOME FINAL COMMENTS

Let me try to summarize the following conclusions:

International Law has promoted a global perspective in relation to biomedical technology. This global perspective has been favored because the State laws lacked ethical and cultural reference points of a clear and undisputed application to the new

---

[7] According to a new text, that was introduced by a reform of the Portuguese Constitution in 1997: "The law shall guarantee the personal dignity and the genetic identity of the human being, specifically in the creation, development and utilization of the technologies and in the genetic experimentation" (art. 26.3).

[8] See Greek Constitution, as modified in 2001 (art. 5.5): "All persons are entitled to the protection of their health and of their genetic identity. Matters relating to the protection of every person against biomedical interventions shall be specified by law".

[9] See Constitution of the Swiss Confederation, as reviewed in 1999 (art. 119 ff.), although there are not set as rights, rather as exclusive jurisdiction of the Confederation. One must reproach, on top of the indefinite legal nature of this constitutional clause, the impropriety of such a detailed and prohibitive regulation – furthermore, adequate of a sanctioning Law –, by all means, excessive in many instances. There is a precedent in former art. 24, as it was added in 1992.

challenges created by the biomedical technology. The UNESCO Declaration on the Human Genome and Human Rights and the Convention on Human Rights and Biomedicine are significant contributions at the international level. Globalization doesn't mean automatically to share cultural values and traditions. We need go further to reach a transcultural ethics and law in the field of human genetics and biotechnology, this is to say, to share a minimum common of those.

This means that there are some values (human rights, juridical goods) that are related to human life sciences that have achieved a universal recognition with great ease. These have not always found an adequate match with certain cultural and ethical conceptions of some non-western human communities.

However, we also find in relation to human genetics and biotechnology a 'soft law' and some 'soft values', non-desirable, though inevitable in some cases, which could have given rise to:

a) Several value contradictions in national legislations, as is happening in the legal framework on the possibility of researching with the human embryo, in respect to which we can simultaneously notice an intense protection and evident lack of protection;
b) a perspective of globalization that is not transculturally sustained, e.g. in relation to the prohibition without exceptions of human reproductive cloning in all the States that have legally undertaken the matter; and
c) the recourse to a symbolic Law, in reference to which the legislator is more worried about expressing a moral and social rejection towards certain activities and to calm society than for an effective legal persecution of the same (as some are still in the future).

Human rights continue to be an unavoidable reference point to better capture the multiple challenges of biomedical technology, in that the development of these rights is the result of an ethical construction that give them the conceptual support and the axiological credibility. In the future, it shall be necessary to look further into the following aspects:

a) Human rights also present an objective dimension that permits through them the protection of realities or situations independent of the possibility of accepting the existence of a subject titleholder of a specific right.
b) Human rights are not only individual rights, but at the same time have a collective dimension that must be implemented, as it serves to guarantee the adequate protection of specific social groups and communities, without decreasing the individual dimension as a hypothetical side effect; and
c) From this point of view, in the future there should be a recognition of the greater importance that shall be granted to social human rights or those that guarantee the coexistence. In order to develop them, we must take into account principles such as responsibility, solidarity, justice, equity, tolerance, non-discrimination and responsibility towards future generations, whatever they could mean.

The implementation of the Universal Declaration on Bioethics and Human Rights of UNESCO should take into account these type of reflections, as a condition necessary, but probably not sufficient, to achieve a universal acceptance.

REFERENCES

Canguilhem G (1966) Le normal et le pathologique. Presses Universitaires de France, Paris.

Cortina A (2004) Una ética transnacional de la corresponsabilidad. In: Serrano V (ed) Ética y Globalización. Biblioteca Nueva, Madrid.

Delmas-Marty M (1997) Le Droit est-il universalisable? In: Changeux (dir) Une même éthique pour tous? Comité Consultatif National d'Éthique, Ëditions Odile Jacob, Paris.

Eser A, Koch H-G (2003) Forschung mit humanen embryonalen Stammzellen. Strafrechtliche Grundlagen und Grenzen Rechtsgutachten von Hans Dahls/Bernd Müssig und Albin Eser/Hans-Georg Koch, Standpunkte Deutsche Forschungsgemeinschaft, Wiley-VCH.

Fagot-Largeault A (1985) L'homme bio-éthique. Pour une déontologie de la recherche sur le vivant. Maloine, Paris.

Grupo de expertos sobre Bioética y Clonación (1999) Report on Cloning, FCS, ed. Siete Calles, Madrid.

Harris J (2004) On cloning. Routledge, London.

Jonas H (1984) Das Prinzip Verantwortung. Versuch einer Ethik für die technologische Zivilisation. Frankfurt am Main.

Kaufmann Art (1999) Rechtsphilosophie, 2. Aufl., C. H. Beck, München.

Rawls J (1971) A theory of justice. Harvard University Press, Cambridge, MA.

Romeo-Casabona C (2004) Laying the foundations for a transcultural Biotechnology Law (Editorial). Law Hum Genome Rev 20.

Romeo-Casabona C (2002) Embryonic stem cell research and therapy: the need for a common European legal framework. Bioethics (Special issue) 16(6).

Romeo-Casabona C (2002) Los genes y sus leyes. El Derecho ante el genoma humano, Cátedra Interuniversitaria Fundación BBVA – Diputación Foral de Bizkaia de Derecho y Genoma Humano & Editorial Comares, Bilbao-Granada.

Romeo Casabona C (1997) Legal limits on research and its results? The cloning paradigm. Law Hum Genome Rev 6.

Saada-Gendron C (1999) La tolerance. Flammarion, Paris.

Singer P (2004) One world. The ethics of globalization, 2nd edn. Yale University Press, New Haven & London.

Sulston J (2004) Society and human genome. Law and the Human Genome Review 20.

Vallespín J (2004) El problema de la fundamentación de una ética global. In: Serrano V (ed) Ética y Globalización, Biblioteca Nueva, Madrid.

UNESCO (2005) International bioethics committee. Universal Declaration on Bioethics and Human Rights, Paris.

UNESCO (2003) International declaration on human genetic data.

CHAPTER 13

GENETICS AND SOCIETY: A DIFFERENT VIEW

JUAN MANUEL TORRES

Centro de Filosofia das Ciências, Universidade de Lisboa, Lisboa, Portugal; Universidad Nacional del Sur, Bahía Blanca, Argentina, jmvtorres@yahoo.com.ar

Abstract: Some questions that could be asked regarding the revolution of genetics would be the following: what is its real influence on the conceptual network of the social sciences, if any? Are there in the scientific network, social or cultural concepts whose meanings have been modified substantially by this revolution? And, if so, to what extent? A first step in order to answer these questions is to determine what is meant by 'revolution of genetics'. Once this is answered by the identification of some knowledges and technologies characterizing this scientific phenomenon, we shall demonstrate that the effect of the revolution of genetics on other disciplines – such as the theory of health – causes a conceptual change that should be considered as a Kuhnean shift in a certain way. In this article we explore this connection in a way that differs from other approaches that usually consider the relations between genetics and society in terms of eugenic policies or those of the geneticization process, by focusing on the impact that genetic technologies have on the notions of health and unhealth

1. INTRODUCTION

When speaking of scientific revolutions, it is unavoidable to mention Thomas Kuhn. In his influential work *The Structure of Scientific Revolutions* (1962) and others, he described the characteristics of a revolutionary shift in science and also its consequences. Since then, the expression 'scientific revolution' has been linked to the meaning that he has given to it. Therefore, we should find out –albeit briefly– whether the revolution of genetics constitutes a case of Kuhnean shift or whether it promotes some changes of this kind in other fields of knowledge. Naturally, in order to do that it is necessary to begin by determining what is meant by 'revolution of genetics'. One strategy to succeed in this task is to identify the genetic technologies that are usually considered revolutionary and, after that, to reach those scientific knowledges that make them possible, at least, in the sense that the former can explain the latter. In the analyses of genetic technologies and in order to illustrate

A. Fagot-Largeault et al. (eds.), The Influence of Genetics on Contemporary Thinking, 195–206.

them, it will be necessary to circumscribe a specific application field. In this sense, we have chosen the medical sciences that, in turn, will allow us to demonstrate the revolutionary shift that actual genetics is producing in the theory of health.

What are the main technological breakthroughs in direct relation to the tools provided by genetic engineering? We think that, basically, there are two. First, our capacity to identify genes and gene segments in our genome and the genome of other species, to know their nucleotide sequences and, finally, their biological meanings – that is, what they are for. Second, our capacity to create and insert genetic sequences in organisms, in such a way that their cell machinery shall produce what these segments codifies. In the realm of medicine – our field of application – these two great achievements are represented by genetic testing and gene therapy respectively. In order to make this contribution self-contained, we will explain the nature and possibilities of both technologies, that due to their contribution to human health are sometimes called 'genetic services'. Then, we will analyze whether those achievements that made these technologies possible meant a rupture with the tradition in biology and biochemistry or, on the contrary, whether they were consistent with traditionally accepted principles.

2. GENETIC SERVICES

Though its strict definition is much wider since it also involves the analysis of chromosomes, proteins, and metabolites (The Report of the Task Force on Genetic, The Report of the Task Force on Genetic, Testing 1999), it is usually considered that genetic testing is the tool that allows knowing the existence of certain RNA or DNA segments in our organism through the identification of their nucleotide sequences. It is an instrument for detection that looks for the specific genetic sequences for which it was designed. With regard to medical sciences, these sequences are normally associated to pathologies, this being the reason why they are called 'deleterious mutations'. It should be noted that a mutation associated to a disease is far from indicating that it constitutes its sufficient cause because it could be just one of the factors for its onset. Nowadays, we know many genetic mutations that are associated to different types of cancer but, in general, they only constitute conditions for their onset. This is the case of many – but not all – abnormalities in BRCA1 and BRCA2 genes, which help the body to suppress the development of cancer, when working properly. Some mutations of these genes are strongly related to breast and ovarian cancer, as well as colon, prostate and other non-gynecological types (MDH, 1998). However, all of us know the difference between being a factor and being a sufficient cause.

In order to emphasize the mistake in the popular formula *deleterious mutation → illness*, it should be noted that even if a mutant sequence were sufficient cause for pathology, this would not mean automatically that its carrier will suffer from it. Together with the deleterious mutations, the Mendelian phenomena of dominance and recessiveness should be considered. Despite of this and other warnings, today it is common to see in the media that the notion of genetic disease is misconstrued, as if it were invariably monogenetic and dominant. Because of the importance given in this contribution to the influence of genetics on some concepts with deep social implications, we point out that this misrepresented notion of *genetic disease* is part of the beliefs which characterize the so called 'geneticization process'.

Perhaps only with the exception of the common form of sickle cell anemia that is caused by a single specific mutation, a genetic test never detects the *mutation* that is the cause of or that is associated to a disease. In other words, the formula *having the test for the mutation associated to x disease → having the test for x disease or for the risk of suffering it* is false, and also part of the beliefs of a society under a geneticization process. Simply because there is not such a thing as *the* mutation that causes a disease, but the *mutated gene* and this usually can be abnormal for *many* mutations. Shortly, there is not one cause for the mutation of the gene responsible for cystic fibrosis, but many (more than six hundred and also hundreds in the case of BRCA1 and BRCA2 genes). Tests find only those mutations for which they were designed to, but don't reveal the others, such as deleterious or non-deleterious ones. This is the reason why a negative testing never allows saying that there are no deleterious mutations for a specific gene, it should be said that no one was found by the test. Anyway, not all mutations have the same frequency in human ethnical groups and, therefore, the same importance.

Though genetic testing is usually associated to the detection of causes and factors related to pathologies, it plays this role in the realm of health sciences. Outside such field, this tool allows identifying individuals and biological relations between individuals, to characterize species, and, sometimes, to measure evolutionary distances among species, organisms, and viruses, as it is the case in the family of immunodeficiency viruses that affect humans and simians (HIV and SIV).

Let's now take a look at some of the advantages offered by genetic testing to medicine. It allows identifying some pathologies when their signs and symptoms are weak or confusing and, in general, when the elements for traditional diagnosing are insufficient, as in the case of acute lymphoblastic and myeloid leukemia, a condition in which it is difficult to discern between alternative diagnoses. In this sense, it can be said that genetic testing is a new way of confirming a classical diagnosis. By the use of genetic testing, it is also possible to determine the presence of viruses – such as HIV and HPV – and bacteria in the organism, even before these invaders begin to carry out their devastating work. A remarkable feature of genetic testing is that it is one of the major tools responsible for a new kind of medicine: the so-called 'predictive medicine', based on the principle that states that genes can indicate disease before any symptoms appear. Thus, by the use of this tool it is possible to detect mutations that are known risk factors for serious pathologies, a fact that allows taking preventive measures in order to delay the onset of diseases or to try to avoid them when possible. We can also mention diseases such as *multiple endocrine neoplasia type 2* (MEN-2) or *hereditary hemochromatosis*, for which genetic testing provides great clinical benefits since it indicates the need for treatment before clinical complications occur (Burke, 2002). By working directly on the genetic level, genetic testing allows to detect individuals who are mere carriers of deleterious genes. In this way, it is possible to identify couples at risk of having a child with a genetic disorder. Predicting that the newborn will suffer health problems allows, in turn, anticipating medical measures, special attention, and dedication.

Finally, there are some genetic tests that allow to detect mutations that in a later stage in the individual's life – adolescence or adulthood –, will become sufficient causes for high risk or fatal pathologies, although the individual with genomes positive for these tests, is physiologically well at the moment. This is the case of Huntington's disease, an autosomal dominant condition that causes progressive motor and dysfunctional disability starting in midlife. With regard to one of the purposes of this contribution – i.e. to demonstrate the revolutionary influence of genetics on the health theory –, it is important to emphasize the existence of this kind of test, that in an informal way we may call a 'time-bomb detection test' (TDT), though its usual name is 'test of asymptomatic individuals'. The name proposed comes from the fact that it is related to those mutations that for obvious reasons should be called 'time-bombs'. Though it is true that only a few time-bomb mutations have been identified up to now, this is because we are at the initial stage in the research of the relations between deleterious genes and diseases and, therefore, in determining what a healthy genome is and what is not. Against the arguments that try to minimize the growing role of predictive medicine it can be said:

> A widespread reason for rejecting expressions such as "genetic health" or "healthy genome" is to consider as paradigmatic of genetic diseases those pathologies whose onset is tied to extra genetic factors, where these factors play roles that are sometimes more important than the genetic basis itself. In other words, we are talking about those scholars who speak of health and genetics taking into account situations in which it is possible to avoid pathologic conditions by modifying lifestyles or other actions. But those, who choose these examples as their favorite, in an attempt to reject any genetic determinism in connection with human illnesses, seem to ignore something that is very common to see in any hospital: pathologies caused by defects in genetic instructions, in which lifestyles or personal histories are absolutely irrelevant.
>
> (Torres, 2006)

Nevertheless, and in order to end our exposition of some genetic testing features, it should be clear that a genetic test is not a disease test, but a tool to determine genetic contribution to the risks of disease, though sometimes these risks may be close to 100%.

Whereas 20 years ago the number of genetic tests could be counted with the fingers of one hand, today we have hundreds of them – one thousand according to the latest estimations – and this number is continuously increasing. It is in this context that the importance of the Human Genome Project can be appreciated in connection with health sciences. Its results allow accelerating the task of sequencing mutant segments, once they have been mapped by cytological studies performed on family members suffering from typical conditions. However, the number of genetic tests tells nothing by itself if the number of genetic diseases is not taken into account, a number that some consider to be close to 10,000. However, although the number of deleterious mutations that are causes of disease or that influence their onset is enormous, this number is limited. It means that, from *a practical point of view*, in the near future, we will get to know, at least in general, what a healthy genome is and, therefore, what it is not. By 'healthy genome', we do not mean a genome free of all mutations because there is not such thing, but that genome free of mutations that are sufficient causes of serious pathologies or that may put us at high risk of suffering them. It should be noted that we are speaking from a practical point of view, because new mutations will appear, mainly due to the mistakes made during genetic material replication processes.

Genetic medicine, also known as 'gene therapy' or 'human gene transfer', can be defined as the insertion of genetic material into cells for therapeutic purposes. Though this definition is a little narrow, it embraces most of the protocols being applied in research medical centers. This new branch of medical sciences tries to insert correct genes to replace their respective mutant genes that are unable to instruct the cellular machinery to produce efficient proteins. There are also protocols that try to increase the natural capacities of some of our cells. Therefore, for example, through the *multi drug resistant protocol* (MDR) researchers attempt to increase bone marrow cell defenses against lethal substances, such as taxol and other drugs used for cancer treatment. Needless to say that this protocol that try to increase natural cell capacities, has nothing to do with the construction of super-men because is used in therapeutic contexts (Torres 1997). Unfortunately, here we face again another belief of the geneticized society, one that is even popular among journalists looking for shocking topics. As the MDR protocol, the introduction of RNA anti-sense sequences to avoid HIV replication does not find its place in the classical definition of human gene transfer (i.e. gene substitution medicine) and is useful to demonstrate the many possible applications of this genetic service.

We should point out that gene medicine is still at the research stage. Some protocols may be considered partially successful, such as that for adenosine deaminase, but none of them is ready for massive consumption. Historically, it is interesting to observe that gene medicine began with the success obtained by French Anderson in two patients suffering from adenosine deaminase deficiency. However, after that, gene medicine got into a dark period of doubts and failures from which it is just beginning to get out through promising protocols for rheumatoid arthritis, neuropathic pain, and muscular dystrophy. Anyway, the pioneer work of F. Anderson and others has shown, beyond any doubt, that the principle over which gene medicine lies is correct: the transfer of genetic sequences into human cells can be interpreted by the cell machinery. Even more, in the last few years, the detection of many of the technological problems that prevent us from using this principle at will has constituted a crucial achievement for medical research. Therefore, to asses that in the next few years we will be able to cure or, at least, to offer some relief to many genetic diseases for which there are not treatment up to now is not a mere prediction.

3. KUHNEAN AND NON KUHNEAN REVOLUTIONS

It is common place to say that the 20th century has been characterized by three scientific revolutions: the atomic revolution, the information revolution, and the biotechnological revolution. It should be noted that 'revolution' here means, at least primarily, the way in which these technologies, which depend on their respective basic disciplines, have influenced society, sometimes for better and sometimes for worse. As it is well known, they have increased our well-being, our control over nature, and our life expectation and have turned our planet into a global village. But for us, the key issue is that these revolutions never suppose or are based on principles opposite to those accepted before the arrival of such technologies. Even more, it can be said that these revolutions are the practical consequences of what was known and accepted at theoretical level. In general, these technologies have shown an absolute consistency with the basic principles, which were able to justify – sometimes a posteriori – their success.

The revolution of genetics is usually considered the child of three key scientific achievements: (i) the isolation of the enzyme DNA ligase, capable of providing a molecular glue to join DNA strands and, in this way, to create DNA segments at will (ii) the isolation of the first restriction enzyme, that allows to cut DNA segments at precise positions and, in this way, to know any DNA/RNA nucleotide sequence (1970); and (iii) the transfer of a gene from one organism to another. It is easy to see, even for those who are not familiar with biology, that (ii) and (iii) constituted the technological bases for genetic testing and human gene transfer respectively.

Technological details aside, the point here is that those achievements were the steps expected in a progressive program that began with the identification of genetic substance by O. Avery et al. in 1944, its chemical structure and replication mechanism by J. Watson and F. Crick in 1953, and, finally, its language or genetic code by M. Nirenberg in 1967. Neither the possibility of transferring genetic sequences between organisms – that today has become common practice – nor that of knowing a nucleotide sequence, constituted a revolution in the sense of rupture with the scientific tradition. In other words, these and other achievements were in no way opposite to the theoretical foundations of biochemistry and what we today call 'molecular biology'. Contrarily, as soon as the biochemical principles were set up, these practical possibilities – that today characterize the revolution of genetics – were predicted, even when it was unknown when and how they could become a reality.

Perhaps few texts are so illustrative in showing the continuity between basic and applied science in the field of genetics as this one by M. Nirenberg (1967), that already shows his social and ethical concerns for the potential effects of the future tools:

> My guess is that cells will be programmed with synthetic messages within 25 years... The point that deserves special emphasis is that man will be able to program his own cells long before he will be able to assess adequately the long-term consequences of such alterations, long before he will be able to formulate goals, and long before he could resolve the ethical and moral problems which will be raised.

Theologists P. Ramsey (1965) and K. Rahner (1967) also shared the same concerns about humanity in view of the applications of future genetic technologies, a fact that confirms the continuity between biological and biochemical principles, on one side, and the technologies that years after caused the revolution of genetics. In brief, it can be said that the revolution of genetics was never a surprise for scientists.

In his book, *The Structure of Scientific Revolutions*, Thomas Kuhn coins a definition of scientific revolution that even today has a deep and extended influence on the academic world. As it always happens, the notion coined by Kuhn was not a *de novo* product; however, he was the one who gave it the meaning with which the expression is used by many members of the scientific community, specially, by those working in the field of basic research.

A characteristic of the Kuhnean concept is that it does not aim at describing the impact of science and technology on our life or society. A *Kuhnean* scientific revolution can be characterized from many perspectives and these come from the many descriptions that Kuhn made of these extraordinary events in his foundational book. Among other things, a revolution implies a dramatic change in the beliefs of the some (leaders) members of a scientific community with regard to a pair of theories – that is, the new one and the one displaced by the first. Scientists adopt new principles because the old ones were unable to explain important phenomena that for this reason were considered to be *anomalies*. Up to here the Kuhnean concept does not seem to be too revolutionary. But there is an additional element: the theories, the old and the new one, are incommensurable between each other and, therefore, cannot be compared, at least, in major aspects. This happens because, although both made use of the same fundamental terms, each one assigns different meaning to them, at least in part. Kuhn's argument aimed at joining the phenomena of meaning change and incommensurability, and runs in the following way: scientists *know* the world in part through the meaning of the terms that appear in their laws and principles. Therefore, if the meanings change, then also the way in which the world is given to us changes. After that, Kuhn's famous and scandalizing words cannot be a surprise: "After a revolution, scientists are responding to a different world". However, in this way, irrationality settles in the history of human reason: there is no progress in science, but alternation in the scientific visions of the world.

The strong criticisms received during the '60s and the maturity of his own ideas leads Kuhn to reformulate and refine his point of view on incommensurability. Although in his book published in 1962 the different ways of perceiving the world were attributed to influences coming from the different meanings that the same terms had in the two theories – the so called 'phenomenon of theoretical load', – in the 70s we witnessed a new concept of incommensurability, one that was freer from psychological influences. Now, the reformulated notion does not imply incomparability, that was one of the consequences of incommensurability and the dialog between scientists of both theories is possible. Nevertheless, there is still an important difference in the perception of reality by each group: the way in which supporters of each doctrine organize their field of knowledge. In a few words, the different taxonomies are now the roots of communication problems among them because scientists of competing theories will keep observing the world in a partial and different way. Among his examples, Kuhn gives one that is especially adequate for our purposes because it is based on theories that do not use metrical notions, but qualitative ones. For the Ptolemaic theory, the term 'planet' denotes, among other bodies, the Sun and the Moon, whereas the Earth falls under another category. In the Copernican view, the Earth becomes a planet, the Sun is now a star and the Moon a new kind of entity: a satellite (Kuhn, 1997, 1981).

It is evident from our explanation that the revolution of genetics does not resemble a Kuhnean revolution. Firstly, the revolution of genetics takes place in the realm of applied sciences. Secondly, and more importantly, the foundations of genetic technology are absolutely consistent with the biological, biochemical, and molecular principles introduced in the mid of the 20th century. This should be understood in the sense that those technologies can be explained from these principles. At this point, our presentation of Kuhn's doctrine appears to have been a waste of space. However, though the revolution of genetics is not a phenomenon as those described by Kuhn, this does not imply at any rate that this revolution could not produce a phenomenon of this kind in a different field of knowledge. It is something that we will demonstrate here and for which we will need to make use of Kuhnean analysis. In this connection, we should remember that a revolution in the Kuhnean sense should be relativized to a discipline and not to the science in general.

4. INFLUENCE OF GENETICS ON HEALTH THEORY

The scientific changes that match the Kuhnean model of taxonomic rearrangements should be determined by historical research. But the point for us is that this kind of shift illustrates in general what is happening nowadays in health theory by the influence of the revolution of genetics, more specifically, by the influence of genetic services. Before seeing why it occurs, we should take a look at the health theory (HT).

The HT can be characterized as the study that aims at determining the meaning of some specific categories. Among these, we can mention, as the most important ones, the categories of *health, unhealth, disease, illness, disability, handicap,* and *injury.* A very important question to be taken into account is that the HT is not medicine or part of it, although both are closely related. In this regard, it is enough to see that there is great disagreement throughout the medical community on the meaning of being healthy or unhealthy, especially in the case of the so-called 'mental diseases' and deviate behaviors. Even more, as we shall see, whereas medicine is going through a period of normal science, the HT is suffering a revolutionary change. In general, we should say that the HT is an interdisciplinary construction in which contributions coming mostly from medicine, biology, and psychology concentrate, but contributions are also coming from anthropology, sociology, and philosophy. In this connection and making even more evident that the HT is not medicine, it should be noted that two of the most significant formulations of the health theory come from philosophers: the biostatistical theory by C. Boorse (1977, 1977) and the finalist theory by L. Nordenfelt (1986, 1995). Regarding the purposes of this contribution, one thing that should be taken into account is that health categories are of great social relevance. To characterize someone as healthy, ill, disabled, etc., always has strong social, economic, and legal effects. So, the application of these categories can be a cause of discrimination or protection.

A characteristic of Boorse's HT is that for him the categories of unhealth and disease are just the same or, at least, have the same extent. In other words, according to his formulation: *if someone is diseased, then he/she is unhealthy*; and *if someone is unhealthy, then he/she is diseased*. In addition, the view that L. Reznek has called 'the medical paradigm' (1991), because of the fact that it is shared by the medical community, is in agreement with this identification of unhealth and disease. This equivalence also has great acceptance with ordinary individuals, something of particular importance for the great social implications of these categories. On the other hand, what is apparently more reasonable than to think that if someone is unhealthy, it is because he/she has a certain disease? It is true that not all scholars agree with this identification, especially those who make a difference between 'disease' and 'illness', which does exist in some contexts in English. So, for L. Nordenfelt, the main contrast lies between health and illness, because for him a disease does not necessarily involve a health problem, if health is defined – as he defines it – as the ability to reach vital goals. Nevertheless, for our purposes, this difference between Boorse and Nordenfelt is not relevant. Another characteristic of Boorse theory, one also shared with other views, is the following: the categories of health and disease (illness in the case of Nordenfelt) are the fundamental ones in the HT. It means, among other things, that they are located at the top of the taxonomy of health categories.

Now, let us consider an individual who has a positive result for a pre-symptomatic test – i.e. for a test able to discover lethal mutations that in the future will be sufficient causes of serious illnesses and that we have informally called 'time-bomb detection test'. For example, the tests for detecting the mutations related to Huntington disease or familial adenomatous polyposis (FAP). Now, two questions should follow: (a) is this individual ill or diseased? (b) is this individual healthy? With regard to the first, the answer is, of course, not. That is so because, fortunately, all health theory schools and also ordinary individuals apply the term 'ill' or even 'diseased' only to those having some kind of organic problems or behavioral disorders, at least to the extent that the latter are consequences of physiological abnormalities. Certainly, there exists an old discussion on whether purely mental diseases exist or not, but we have to leave this question aside. Now, we go to the second question, is he/she healthy? If we accept the prevailing view in HT, for example Boorse view, the medical paradigm, and also the opinion of ordinary individuals, then we should conclude that he/she is indeed healthy. The rationale is simple: they oppose contradictorily *health* and *disease*, therefore, *tertium non datur*. Nevertheless, we think that not all of us would accept to call an individual 'healthy', whose life will be interrupted early – for example in his/her adulthood – because of his/her peculiar genome and not because of external circumstances.

If we had an uncontroversial definition of health, one shared by all schools working in HT, then our question would be relatively easy to solve. Unfortunately, this is not the case because one of the main issues in the epistemology of medical and health sciences remains to be solved: what health is. However, we do not need to wait for a complete answer about the nature of health in order to solve our problem, that is, whether an individual should be considered healthy, if she/he is positive for a pre-symptomatic genetic test related to serious pathologies. In this sense, our argumentation goes in the following way: although there are many and important differences in the definitions of health provided by the different views, this circumstance does not exclude that there can also be some important coincidences. In fact, there are some agreements on it, that play the role of necessary – though not sufficient – conditions for health, as, for example, the condition to be able to move by oneself. Among these necessary conditions, we have identified one that we have called the *basic biological requirement* (Torres, 2002). It consists in the possibility of going through the three stages that characterize human life and the life of some other species: childhood, adulthood, and old age. It is on the basis of this requirement that we usually project our future because, as intelligent beings, we have the capacity – at least in principle – to determine our own destiny by making decisions; typically by assuming long-term responsibilities, as getting married or having children. It should be clear that the basic biological requirement is not equivalent to health nor does it constitute its definition. Simply, this requirement is an obviously necessary condition for it. If we accept this – and we must –, then it is evident why an individual with a genome that contains a time-bomb mutation, cannot be considered healthy: she or he will not be able to go through the three stages that are connatural to human life.

An immediate consequence from the last paragraph is that although those who are ill are unhealthy, not all unhealthy individuals are ill (or diseased). Certainly not those with a positive result for those tests able to detect mutations related to future serious pathologies. Of course, there could be some individuals with low aspirations. For example, those for whom it is not important to have such a tragic prognostic or to die prematurely. However, as rightly pointed out by Nordenfelt (2001, 67), to take into account low aspirations in order to define health fails for many reasons. A second consequence is that the up to now accepted taxonomy of health categories should be changed. After the arguments presented, it is clear that the top disjunction is no longer *health* v . *disease* (or illness), but *health* v . *unhealth*. Two considerations about it: (i) Certainly, *health vs. unhealth* involves a truism or tautology, at least if we use standard bivalent logic. However, it is not the same in the case of our affirmation that disease is only a subset of unhealth, which is wider than the first because it also includes individuals without *actual* organic or behavioral problems, but future ones depending on their genome. (ii) This result that affects the taxonomy of health concepts is a consequence of the revolution of genetics, since genetic testing is a possibility provided by this revolution. Certainly, before the appearance of genetic testing we knew a lot about genes, but we had no direct access to them, an access that allowed for detecting specific mutations or viral genomes. By that time, our approach was through the effects of genes on the organism, which means – in general –, when people were already symptomatic and, therefore, suffered from the condition.

"After a revolution, scientists are responding to a different world", in this way Kuhn attempted to characterize these exceptional shifts. Accordingly, it proceeds to ask now: how different is the world in which scientists are living today? Before we answer, we should emphasize that we do not claim that the shift operated in health theory is a revolution in the *full Kuhnean sense*, rather that this movement presents a feature – albeit conspicuous – of what Kuhn himself called 'scientific revolution'. The world we are talking about has already become partially evident by predictive medicine, at least in part. We move towards a world in which a patient's individual genome will be part of his/her medical record. However, we should be careful because one thing is a revolution in health theory and another very different one is a revolution in medical sciences. As it was said, medical sciences are going through a normal period, but it is not the same as with the health theory, in which the category of health is today becoming more and more demanding and restricted, whereas that of unhealth becomes wider. In turn, it is clear that this change in health theory is bound to influence health policies in developed countries. Individuals with pre-symptomatic positive tests related to serious pathologies, that is, people who lack the *basic biological requirement*, will have special rights and protection. As an example, special retirement plans should be designed because many of the current ones are thought for a stage in life to which such individuals will hardly reach (Torres, 2006).

5. CONCLUSIONS

Genetic make-up, organism, and behavior are the three dimensions in which individuals are considered and treated by medical sciences nowadays. However, whereas the last two have long been known to medicine, the genetic dimension is quite new to it because of our relatively recent direct access to genetic make-up. This last approach also explains well why in traditional formulations of health theory the genetics revolution and its influence on health categories are absent.

Philosophers and social thinkers have dealt with the effects of the genetic revolution from many perspectives. So, J. Derrida (2002) focuses on genetic breakthroughs in connection with eugenic practices that would result in a world of superhuman and sub-human beings, whereas Baudrillard (2000) expresses his concerns about human cloning which would make human beings immortal. Habermas (2003) also addresses the problems of eugenic practices and their consequences for the self-perception of those whose genetic make-up has been programmed – typically by parents –, i.e. existential conflicts arising from the knowledge of being the result of a design. A society of gene-rich and gene-poor individuals caused by the buying and selling of genes is for Fukuyama (2002) a probable scene for the years to come.

Unlike in the approaches mentioned above and in others referring to eugenic manipulations, which sometimes are very far from actual technological possibilities, we have considered the consequences of genetic revolutions by focusing on genetic services and their consequences on health theory. As we have extensively demonstrated in this work, one of these consequences is a substantial change in the meaning of the traditional categories of health and unhealth. This is a theoretical movement already perceived by important experts in philosophy of medical sciences (Bayertz, 1997; ten Have, 2001).

REFERENCES

Baudrillard J (2000) The vital illusion. Columbia University Press, New York.

Bayertz K (1997) What is special about molecular genetic diagnostics? The J Med Philos 23:247–254.

Boorse C (1977) Health as a theoretical concept. Philos Sci 44:542–573.

Boorse C (1997) A rebuttal on health. In: Humber J, Almeder R (eds) What is disease? Human Press, Totowa.

Burke W (2002) Genetic testing. The N Engl J Med 347 (23):1867–1875.

Derrida J (2002) The aforementioned so-called human genome. In: Rottenberg E (ed) Negotiations. Stanford University Press Stanford.

Habermas J (2003) The future of human nature. Polity Press. Cambridge.

Fukuyama F (2002) Our posthuman future. Consequences of the Biotechnology revolution, Profile, London.

Kuhn T (1962) The structure of scientific revolutions. The Chicago University Press Chicago.

Kuhn T (1970) Reflections on my critics. In: Lakatos I, Musgrave (eds) Criticism and the growth of knowledge. Cambridge University Press.

Kuhn T (1981) What are scientific revolutions. In: Krüger, Daston, Heidelberger (eds) The probabilistic revolution. The MIT Press.

Kuhn T (1997) "The natural and the human sciences" in Hiley, Bohman and Shusterman eds. The Interpretative Turn. Philosophy, Science and Culture, Cornell University Press, 17–24.

MDH (1998) Genetic testing for susceptibility to breast cancer. Minnesota Department of Health June:1–21.

Nirenberg M (1967) Will society be prepared. Science 157:633.

Nordenfelt L (1986) Health and disease: two philosophical perspectives. J. of Epidem and Comm. Health 40:281–284.

Nordenfelt L (1995) On the nature of health, 2nd edn. Kluwer Academic Publishers, Dordrecht.

Nordenfelt L (2001) Health, Science and Ordinary Language. Radopi, Amsterdam – N. York.

Rahner K (1967) Zum Problem der genetischen Manipulation. In: Schriften zur Theologie, Benziger Verlag, Zurich.

Ramsey P (1965) Moral and religious implications of genetic control. In: Roslanski DJ (ed) Genetics and the future of man. North-Holland Publishing, Amsterdam.

Report of the Task Force on Genetic Testing (1999) In: Holtzman N, Watson M (eds) Promoting safe and effective genetic testing in The United State. J. Hopkins University Press, Baltimore.

Reznek L (1991) The philosophical defense of psychiatry. Routledge and Paul Keegan, London.

ten Have H (2001) Genetics and culture. The geneticization thesis. Medicine, Health Care and Philosophy 4:295–304.

Torres JM (1997) On the limits of enhancement in human gene transfer: drawing the line. The J Med Philos 22:34–53.

Torres JM (2002) The importance of genetic services for the theory of health: a basis for an integrating view of health. Medicine, Health Care and Philosophy 5:43–51.

Torres JM (2006) Genetic tools, Kuhnean theoretical shift and the geneticization process. Medicine, Health Care and Philosophy 1:3–12.

NAME INDEX

SUBJECT INDEX

LOGIC, EPISTEMOLOGY, AND THE UNITY OF SCIENCE

Editors:
Shahid Rahman, *University of Lille III, France*
John Symons, *University of Texas at El Paso, U.S.A.*

---

1. S. Rahman, J. Symons, D.M. Gabbay and J.P. van Bendegem (eds.): *Logic, Epislemology, and the Unity of Science*, Vol. 1. 2004 ISBN 1-4020-2807-5
2. D. Vanderveken (ed.): *Logic, Thought and Action*, Vol. 2. 2005 ISBN 1-4020-2616-1
3. J. van Benthem, G. Heinzmann, M. Rebuschi and H. Visser (eds.): *The Age of Alternative Logics*. Assessing Philosophy of Logic and Mathematics Today, Vol. 3. 2006 ISBN 1-4020-5011-9
4. J. Faye, P. Needham, U. Scheffler and M. Urchs (eds.): *Nature's Principles*, Vol. 4. 2005 ISBN 1-4020-3257-9
5. B. van Kerkhove and J.P. van Bendegem (eds.): *Perspectives on Mathematical Practices*. Bringing Together Philosophy of Mathematics, Sociology of Mathematics, and Mathematics Education, Vol. 5. 2006 ISBN 1-4020-5033-X
6. A. Fagot-Largeault, S. Rahman and J.M. Torres (eds.): *The Influence of Genetics on Contemporary Thinking*, Vol. 6. 2007 ISBN 978-1-4020-5663-5

---

springer.com

Kolibri 360 GmbH
Ferdinand-Jühlke-Straße 7
99095 Erfurt, Deutschland
produktsicherheit@kolibri360.de